PRAISE FOR

Ambling into History

"Honest and perceptive." —*The New Yorker*

"Bruni has [written] a classic American story of a man who had to reach inside himself to find a president. Draw up a stool and come along for a great ride." —Richard Ben Cramer, author of *What It Takes: The Way to the White House*

"A judicious portrait of our forty-third president. . . . [Bruni's] astute analysis of Mr. Bush as a particularly American character . . . will become a starting point for future biographers." —Robert Dallek, *New York Times*

"The equivalent of a front-row seat [to] history." —*USA Today*

"Adroitly examines Mr. Bush's quirky personality under the microscope of the long campaign. . . . A human, even affectionate, study." —*Dallas Morning News*

"Everything we've come to know about George W. Bush, and then some, is found in the pages of *Ambling into History.*" —*Washington Times*

"Effectively conveys the mind-numbing falsity of the presidential campaign process. . . . Bruni brings us the real Bush. . . . A great accomplishment." —*The Oregonian*

"An insightful and intriguing book. . . . [It] portrays a man who won the presidency by being engagingly ordinary—and then discovered an amazing ability to lead his nation in an extraordinary crisis." —*Boston Herald*

"The most insightful view of the day-to-day activities of the press covering a presidential campaign . . . since Timothy Crouse's *The Boys on the Bus.*" —*Christian Science Monitor*

"By paying George W. Bush the compliment of close attention—analyzing every gesture and word with the eye of a theater critic and the ear of a skilled reporter—Bruni offers fresh insight into the man behind the disciplined message of his campaign."

—George Stephanopoulos,
author of *All Too Human: A Political Education*

"A perceptive personal look at the impish, seductive, and charmingly ordinary man thrust by history into an extraordinary role. With revealing new anecdotes and a great eye for character, Bruni has Bush's number—for better or for worse."

—Jonathan Alter, *Newsweek* magazine

"During Election 2000, Frank Bruni was the reporter other reporters turned to first for his acute perceptions about George W. Bush, the candidate. Here Bruni gives us an after-hours, over-a-beer distillation of his experiences on the trail and at the White House. His insider's account is both a sardonic indictment of the fallen state of American campaign journalism and an invaluable character study of a president who turned out to matter far more than anyone in the press section of the plane thought he would."

—Jacob Weisberg,
chief political correspondent, *Slate* magazine

"An amazing look at the unlacquered W. slip-sliding toward the Greatest Fame on Earth, while his handlers work to dress him up for the show."

—*Detroit Free Press*

"Bruni gives readers the first private glimpse of Bush behind-the-scenes, a portrait that is sometimes intimate, sometimes skeptical, but always human."

—Jane Mayer,
staff writer for *The New Yorker*
and co-author of *Strange Justice* and *Landslide*

About the Author

FRANK BRUNI, a reporter at the Washington bureau of the *New York Times,* now writes full-time for the *Times* Sunday magazine. For his previous work on other subjects, he was named a finalist for the Pulitzer Prize for feature writing and a winner of the Polk Award for metropolitan reporting. He has appeared on ABC-TV's *Nightline* and other broadcasts to talk about the Bush campaign and presidency.

ALSO BY FRANK BRUNI

Consumer Terrorism
(with Elinor Burkett)

A Gospel of Shame
(with Elinor Burkett)

Ambling into History

THE UNLIKELY ODYSSEY OF GEORGE W. BUSH

FRANK BRUNI

Perennial
An Imprint of HarperCollins*Publishers*

A hardcover edition of this book was published in 2002 by HarperCollins Publishers.

HarperCollins books may be purchased for educational, business, or sales
motional use. For information please write: Special Markets Department,
Collins Publishers Inc., 10 East 53rd Street, New York, NY 10022.

dition published 2003.

Mazzone

taloging-in-Publication Data is available.

10 9 8 7 6 5 4 3 2 1

For my father, Frank Sr., who is the most honest and decent person I know. And for my nieces and nephews (Leslie, Frank, Erica, Sarah, Harrison, Mark and Gavin), who are easily the most adorable.

PROLOGUE

FOR GEORGE W. BUSH, SEPTEMBER 11, 2001, promised to be a pretty typical day. Not entirely typical: he was on the road, in the middle of a presidential visit to Florida, and he usually spent much more time in the White House or at Camp David or even at his ranch in Texas, preferring surroundings he knew well to those he didn't, rituals that were predictable to those that harbored the seeds of chance. But his schedule that morning featured his favorite traveling act, a kind of color-by-numbers event he could do in his sleep. He was going to visit an elementary school in Sarasota, mingle with the kids there, smile at them, listen to them read—nothing even remotely taxing. The pictures would be terrific, and Bush was likely to be relaxed and on top of his modest game. He tried to make sure of that by beginning the day in one of his favorite ways—with a swift, sweaty run. The presidential motorcade drove twenty minutes from the resort where he was staying to a nearby golf course so he could log his miles in a suitably safe, placid setting.

But around 9:05 A.M., everything changed. Bush was at the school, in a classroom, among the children, and he had already been told about a plane hitting one of the twin towers of the World Trade Center in Manhattan—a freak accident, or so it had initially seemed. Now he was being told about a second plane hitting the second tower, and he was being told that it wasn't an accident at all. His chief of staff, Andy Card, leaned toward him and whispered in his ear: "America's under attack." The television cameras and the people all around the president didn't pick up those words, but they saw the reaction on Bush's face, a brief, muted play of shock, wonder and horror. That was the picture, unplanned and unstudied, that would endure.

There was no script for this, and it quickly showed. After retreating from the classroom for little more than fifteen minutes and scribbling some words with a black felt-tip on a yellow pad, Bush reemerged to make his first public statement about the unthinkable events. He looked somewhat stricken. He sounded slightly shaky. It wasn't his voice so much as his words, so jarringly colloquial. He vowed to "hunt down and find those folks" who did this. Then he said he would return to Washington immediately.

But he didn't, not for the next nine hours. Concerned about his safety, national security officials directed him elsewhere, keeping him on the move and on the fly in a manner that could only heighten the nation's nervousness, steering him toward military locations far removed from the capital. At the first of those, Barksdale Air Force Base near Shreveport, Louisiana, Bush spoke again. This time, the language was more polished and ominous. "Freedom itself was attacked this morning," Bush said. "The resolve of our great nation is being tested."

Despite the considerable sweep of that statement, it left something out, something that Bush surely recognized. In a way that neither he nor his advisers could ever have imagined, Bush was also being tested, and it was anyone's guess how he would bear up. Few of his immediate predecessors in the presidency had taken office with such a tenuous claim on widespread popular support and confidence, the legacy of one of the closest and most bitterly disputed elections in American history. Few had entered the White House with such a brief career in public service, such a late-blooming interest in the position, such a spotty body of knowledge and such hurried preparation, inasmuch as there could ever really be any.

Here was a man who had sometimes indicated ambivalence about his ambitions, who had seemed intent, over his first eight months as president, on making his role manageable and shrinking it down to size. Here was a man whose Republican supporters and even a few of his advisers had worried could seem small and insubstantial. He projected affability more easily than authority, levity more readily than gravity. He favored informal banter over formal oratory. All of this had made him an unusual candidate for the presidency in the first place, and all of this had posed questions about how smoothly and successfully he could slip into the role under normal circumstances. With the events of September 11, those questions loomed even larger, as Bush was asked to command and console a nation through a new era of jittery uncertainty and a crisis more challenging than those that many other presidents had faced.

The arc of his life, his political career and his pursuit of the presidency had been one of slowly, although fitfully, deepening seriousness. But was he ready for—and capable of—this? I had been watching him for more than two years, as the first *New York*

Times reporter assigned to his presidential campaign and then as one of the newspaper's White House correspondents. I had traveled tens of thousands of miles—and exchanged at least as many words—with him. But I wasn't at all sure what to expect.

In many regards, the Bush I knew did not seem to be built for what lay ahead. The Bush I knew was part scamp and part bumbler, a timeless fraternity boy and heedless cutup, a weekday gym rat and weekend napster, an adult with an inner child that often brimmed to the surface or burst through. A scene that always stayed with me was an interview I did with him in June of 2000, as we flew to California aboard his campaign plane. The subject was his wife, Laura, because I was putting together a profile of her, and after Bush and I had been talking for about fifteen minutes, the flight attendant walked by with the small, moist, hot towels that were distributed before a meal. Bush took his, cleaned his hands with it, and then did something less expected: draped it over his entire face and turned abruptly toward me. He was, at this point, the presumptive Republican presidential nominee, with a large and seemingly sturdy lead over Vice President Al Gore in all the national polls. And he was playing a toddler's game of peek-a-boo.

He could be surprisingly casual and silly. I met and was first introduced to him one morning shortly after I hit the campaign trail in August of 1999, and by that afternoon, at the next event I attended, he was already shouting out to me in a familiar, we-go-way-back fashion. "Frankie Boy!" he said as he left a community after-school center at which he had addressed one of the administrators, a man he also barely knew, as "Big Al." "Are you ready for At-LANTA?" Bush asked me. This was where he would be heading the following day, and he announced it with the exaggerated, mindless fervor of a college freshman bound for a keg party. Then he eagerly summoned Laura over to meet me, as if we were these two fabulous people in his life who simply *had* to get to know each other.

I missed some passages of his campaign, like his maiden campaign voyage in June of 1999, when he got on his plane's public-address system and greeted the reporters crammed into almost every seat, a turnout indicative of his early, outsize promise as a presidential contender. "Please stow your expectations securely in your overhead bins, as they may shift during the trip and can fall and hurt someone—especially me," Bush said. But I was there, on the final day of his campaign, when he once again took to his plane's public-address system to alert us before a rally in Bentonville, Arkansas, that we were about to get our "last chance for malaprops," which tumbled so reliably from his lips. This was also Bush—lavishly self-deprecating, defiantly proud of his own failings and foibles.

His playfulness, nonchalance and occasional clumsiness survived his transition from the Texas governor's mansion in Austin to the White House. During his first days as president, as he welcomed various groups of lawmakers into the West Wing, he took to calling Representative George Miller, a burly California Democrat, "Big George," and he decided to refer to Representative Fred Upton, a young-faced Michigan Republican, as "Freddie Boy." He did not work on most Saturdays and Sundays, and when he did, traveling to Pennsylvania one weekend morning to attend a retreat of House Democrats, he unabashedly told the press corps that his afternoon back at the White House would include a little shut-eye. He did not take on an air of exaggerated self-importance, but instead sat behind his desk in the Oval Office and lampooned the obsequiousness that so often came his way. "My, Mr. President, you look fabulous today!" he would exclaim. Or, "Mr. President, that was a magnificent speech!"

True to his untamable tongue, he flubbed his lines around reporters, at one point saying, "I am mindful not only of preserving executive powers for myself, but for predecessors as well." He reached for impromptu poetry and came up with spontaneous doggerel, using the occasion of his first event in the Rose Garden to rhapsodize about the setting as "such a beautiful, beautiful part of our national—our national—really our national park system, I guess you would want to call it."

When he stood beside British prime minister Tony Blair at Camp David, a reporter noted that the two men seemed like very different people and asked Bush whether they had any personal interests in common. "Well," Bush said, "we both use Colgate toothpaste." Blair interjected, "They're going to—they're going to wonder how you know that, George!" Bush then alluded to the fact that they both liked physical exercise, ending the news conference with an exhortation. "See you at the gym!" Bush declared.

But there were other sides of Bush, too—sides that provided more heartening omens for his ability, or at least his determination, to meet the demands that were placed on him after September 11. This was apparent at his inauguration on January 20, 2001, when he walked through the open doors of the U.S. Capitol and onto the inaugural stage with the stiffened gait of a new, painfully earnest military recruit intent on fulfilling his charge. In his remarks, he described the American story as one of "a slave-holding society that became a servant of freedom," of a "flawed and fallible people united across the generations by grand and enduring ideals." He asked Americans to seek "a common good beyond your comfort" and implored them to be "citizens, not spectators; citizens, not subjects." As he did so, his voice trembled with conviction, and when he looked toward his father, who had once held the same office, his eyes held the suggestion of tears. Standing before America was Bush the sentimental patriot, a man exhilarated and emboldened, in the most guileless of fashions, by the idea that he was claiming a place in history.

He held onto that wonder and that profound sense of responsibility. During his first months in the presidency, Bush constantly reminded any of the aides who rode with him in Marine One, the helicopter that would deposit him on the White House lawn, to take in the view of the city's monuments as it descended. He had not ceased reveling in the sight himself. On a trip to Tennessee, he told voters: "I'm pinching myself every day. It is an unimaginable honor to represent the great people of this country." He told a group of state lawmakers who visited Washington: "I cannot tell you what an honor it is to be the president and to drive in those cars with the American flag flying and to see people lining the roads, waving to the—waving to the office."

And he sometimes rose to the most important occasions, exhibiting precisely the kind of poise and presence that had so often eluded him and challenging his detractors' assessments of him as the emptiest of suits and lightest of weights. His first address to the gathered members of Congress, in late February of 2001, was delivered cogently and persuasively, an achievement not only of expert speechwriting but also of effective speechmaking, and he went one better by garnishing it with self-consciously charming, cunning gestures that were perfectly made for television. On his way into the chamber of the House, he took an especially long time getting to the microphone; he zigged this way and zagged that way, offering handshakes to the male lawmakers and hugs to the female lawmakers. They beamed, and whatever else

Americans may have come to think as Bush outlined his principles and his agenda over the next forty-nine minutes, many of them had to be feeling that he was a pretty nice guy.

But there was and could be no occasion like the one he confronted when the terrorist attacks on the World Trade Center and the Pentagon killed about three thousand Americans and laid siege to the country's sense of security. Of all people, George W. Bush needed to be the commander-in-chief and the communicator-in-chief and the comforter-in-chief, a national wellspring of strength and solace. And right away, there were signs of the awkward fit. He gave a television address from the Oval Office on the night of September 11 and looked, in that tight close-up, as he had at other tense times, such as the long weeks in November of 2000 when the election remained unsettled and his and Gore's lawyers ricocheted through Florida's courts. There was a too-narrow scrunching of his eyes and too-tight set to his lips, a stab at purposeful seriousness that barely broke the skin. And the words, alternately pretty and matter-of-fact, were too few. He was on and off in under six minutes—chore done, mission unaccomplished.

Two days later, he allowed reporters and a television camera into the Oval Office to record his side of a telephone conversation with New York mayor Rudy Giuliani and New York governor George Pataki, and it was less than inspiring. There were long, uncertain pauses between the beginnings and the ends of some of his sentences, a halting fumbling at the right thing to say. But then he finished the call and fielded a few questions, and that was when he showed glimmers of something different and better. He was asked if he believed that one of the unrealized goals of the terrorists had been to assassinate him, and then about "what kinds of prayers you are thinking and where your heart is for yourself."

"Well," Bush said, "I don't think about myself right now. I think about the families, the children. I am a loving guy." With those words, tears welled in his eyes, but they didn't make him look weak. They made him look compassionate, just like one of his campaign slogans had promised. They signaled to the country that he was not at all detached from the crisis at hand, that the person who would have to lead them through it shared their heart-piercing sorrow. It was only a public moment, as were the ones that failed to flatter him, and all of these could be dismissed as meaninglessly theatrical. But theater is an integral part of the presidency, and Bush, in any case, had never been much of an actor. The expressions that flashed across his face could be trusted as accurate reflections of what he was really thinking and feeling.

Besides, much of what he was actually doing was also right. He fashioned events to underscore his plea that Americans not associate all Muslims with the misguided zealots at the center of the terrorist attacks, a message that was not only moral but strategic: it assuaged the moderate Arab states, which he needed in the international coalition he was building. Appreciating how difficult it might be, in the months ahead, to maintain Americans' support for his stated war against terrorism, he used his public comments, day after day, to tell them that this would not be easy or satisfyingly dramatic and might yield only incremental victories. And while it was surely tempting, politically and emotionally, to launch an immediate military strike, no matter how ineffective, he waited. Then he waited some more.

In the meantime, he walked tall into the national spotlight, which he had once inhabited so tentatively and sparingly. He flew to New York City to see for himself the rubble where the World Trade Center once stood and to tell the rescue workers that their country saluted them. Without a podium at the ready in those dusty streets, he climbed atop a fire truck and grabbed hold of a megaphone that was extended to him.

"I want you all to know that America today—America today is on bended knee in prayer for the people whose lives were lost here, for the workers who work here, for the families who mourn," he said.

"I can't hear you," someone in the crowd shouted back.

And Bush, without missing a beat, offered the perfect response. "I can hear you," he said. "I can hear you. The rest of the world hears you. And the people who knocked these buildings down will hear all of us soon." His aides were so unprepared for this unrehearsed sequence of events that they realized only later that they should have grabbed the megaphone and packed it away for posterity, a historical prop to be displayed someday in a place like a presidential library. Over the next week, they would try, fruitlessly, to find it.

That weekend, at Camp David, Bush decided to raise his voice another notch. Although some aides initially resisted—wondering what he could really say and, perhaps, whether such a grand and sober setting was his best showcase—Bush wanted to address Congress in the House chamber, and he wanted to do it soon. On Monday morning, he ordered up a first draft of a speech, and he set a deadline of Monday night. Most of these words would not be his own, but it would be up to him to make them sound as they should, to convince the country that he knew what he was doing and where he was heading. After several drafts, the remarks were ready, and so was he. On the day of the big speech, September 20, a cap on one of his back teeth fell off, which was bound to cause him some discomfort, but he chose to disregard it. He reread his remarks. He took a late-afternoon nap. He had dinner with Tony Blair. And that night, he faced the nation.

"We are a country awakened to danger and called to defend freedom," he told the hundreds of lawmakers seated below him and the tens of millions of Americans riveted to their television sets. "Our grief has turned to anger and anger to resolution. Whether we bring our enemies to justice or bring justice to our enemies, justice will be done." He asked Americans to press on and "to live your lives and hug your children." He promised them that "this country will define our times, not be defined by them."

"As long as the United States of America is determined and strong, this will not be an age of terror," he said. "This will be an age of liberty here and across the world. Great harm has been done to us. And in our grief and anger, we have found our mission and our moment."

Had Bush found his? His demeanor that night suggested so. It had none of the squinty, squeaky hyper-earnestness that had undercut his effectiveness in the past. It conveyed genuine emotion, not a tensed-up facsimile of feeling. There were no flubs, even if he did persist in depriving "terror" of its final consonant, as if it rhymed with Scarlett's fabled Southern manse as well as her last name. And people close to him, even those less practiced at spin, said that he felt, after a life initially filled with false starts and sloppy behavior and as much mediocrity as excellence, that he had inherited his true purpose, the task by which he would be judged and defined.

On the phone with me shortly after Bush's address to Congress, one of his lifelong friends, Joe O'Neill, mused, "What is it they say? 'Some people are born great, some people grow to greatness and some people have greatness thrust upon them.' That's a paraphrase. With George, it's a combination." He explained that Bush was born great in the sense that he was the son of a wealthy, talented man who became the president: Bush had a propitious set of bearings, a better opportunity than most people to pursue big things. "He did have that leg up," O'Neill said. But Bush didn't maximize it for a good long while. Then, O'Neill added, "After his years in the wilderness, he put together a package that was attractive enough to get him elected governor, and he did a damned good job. He grew in that job." Even so, it wasn't a seat of greatness; it never confronted him with a test on that kind of scale. "Now," O'Neill concluded, "we have a situation like this that really requires and demands a leader."

Bush had his own style of leadership, his idiosyncratic ways of looking at and filtering the world. When he told Senator Chuck Schumer, a New York Democrat, of his concern about the country staying patient with a sustained war against terrorism, the point of reference he used had to do with professional sports—namely, his beloved baseball. Bush said he didn't want Americans to "lose interest when the World Series rolls around." He also told visitors to the Oval Office that he wasn't going to let these terrorists "steal home on me."

When he called a friend several weeks after September 11 to offer assurances that he was doing well, he offered as evidence the fact that he had recently run his customary three miles in twenty-one minutes and thirty seconds, a pace of seven minutes and ten seconds per mile. That was faster than Bush had been doing in recent years, and deliberately so, and it grew faster still: he later boasted to the people around him that he had clocked a mile in under seven minutes. Bush was proving to himself, in his own fashion, how strong he could be. He was recasting the demands on him to exceed expectations in terms of a more familiar, tidier trial.

His eloquence sometimes faltered as he stammered into folksy locutions and stumbled into outright gaffes. Sounding like a Western sheriff, he said that he wanted the terrorist mastermind Osama bin Laden "dead or alive." Sounding like a commando, he kept talking about "smoking" the terrorists in Afghanistan out of their caves. In a speech at the CIA headquarters, he used the non-word "misunderestimated" three times in one paragraph, spoke of a "war that declares a new declaration" and talked about what was necessary "to conduct a winning victory." In a speech at Reagan National Airport, which was finally reopening more than two weeks after the terrorist attacks, he said that "ticket counters and airplanes will fly out of Reagan airport." Two days later, addressing federal workers at the Department of Labor, Bush made clear that "there's no doubt in my mind—not one doubt—that we will fail."

Even some of his prominent Republican supporters worried about these recurring mistakes. "In some ways, they're endearing," one of them told me. "But every time he speaks these days, he's commanding the world stage. Everybody's looking at his leadership. Everybody's looking at his judgment." And everybody was looking when, in a visit to the Federal Emergency Management Agency, he began his remarks in a vein every bit as jocular as he ever had. After being introduced by the administration's director, Joe Allbaugh, a longtime friend with a beefy build and a bizarre military-style haircut, Bush said, "I admit, he's not pretty to look at."

I focus on small moments like this because I believe they reveal every bit as much about Bush as large ones and capture what made him so fascinating. That is my prejudice throughout this book, which is dedicated primarily to what Bush looked and acted like on the edges of what was usually considered news, to the personality behind the policies and the often offbeat character that flickered through the frippery and pomp. The nature of his conservatism, nuances of his proposals and contours of his biography have been fairly well established. While this book occasionally revisits those details and freshly appraises them, it doesn't dwell there for very long.

And it is set in large measure during his campaign for the presidency, because that was the last time when reporters like me had a degree of access to him that allowed for privileged glimpses of what made him tick and made him mad, what he relished and did not. Once he entered the White House, the moat between him and us grew broader and deeper, as it invariably does when any man becomes president. We watched Bush from a distance, and we saw him less keenly.

One of the most interesting stories of his presidency was simply a higher-stakes continuation of one of the most interesting stories of his campaign, and what he did after September 11 was what he had done all during his unlikely odyssey to the White House. He caromed between who he needed to be and who he was, between the lofty duties and trappings of where he was going and the humbler habits and dimensions of where he had been, sometimes rejecting one or the other, forever seeking a tenable middle ground between the two. Evaluating his progress was tricky, due partly to the stunning superficiality of American politics, which is another subject of the following pages. That superficiality often filled me with doubts about whether the process by which we elected presidents assured us of getting the right person for the job. That superficiality also extended to the oversimplification and groupthink in the news media's coverage of campaigns and political figures, which this book also examines. When Bush seemed lost, it was sometimes because journalists preferred that narrative and fell, one after another, into a conformist line; when he seemed found, it was sometimes because that was the best next chapter, and those of us chronicling him were eager to turn the page. The truth was less easily pigeonholed. As, in the end, was Bush.

CHAPTER ONE

IT WOULD BE HARD TO IMAGINE AN EVENT more somber than the one Bush attended one Sunday morning in September of 1999 at Texas Christian University in Fort Worth. Although the outdoor stadium he entered was usually a place of frenetic activity, it was now a scene of eerie stillness and quiet, its thousands of occupants sitting or standing with their heads bowed. They had come not to cheer a team but to mourn a tragedy, the deaths of seven people inside a nearby church. Once again, someone with a grudge and a gun had vented his anger with bullets, and once again, a nation reeled from the senselessness of it all.

Just a few days earlier, parishioners at a prayer service at the Wedgwood Baptist Church had been singing a pop version of a traditional Christian hymn when a long-haired man in jeans and a black jacket barged in, screaming obscenities. He opened fire, hitting more than a dozen of them, including many teenagers, before he was finished and turned his weapon on himself. It was the deadliest shooting in the city's history, and it precipitated the anguished and inevitable questions about what could have been done to prevent the violence, about why the hail of gunfire that traced back to San Ysidro and proceeded through Columbine continued to go on and on. Bush found himself in an especially awkward position, because his record as Texas governor was one of defending gun owners' rights; he had even signed a law permitting private citizens with proper licenses to carry concealed weapons. Now he, too, was under a different kind of fire.

The public memorial service at the university wasn't going to change that. But it gave him an opportunity to signal, through his presence, that he was not insensitive to the bloodshed, that he cared. It promised to cast him in the responsible, nurturing light of a leader come to comfort those he led, and because he had recently begun his presidential campaign, it was guaranteed to bring him national attention. He had made the judicious decision not to speak—and, thus, not to make his appearance seem overtly political—but there was a prime, center row of seats for him and his intimates. Print reporters, including me, positioned ourselves as close to it as we could.

Bush saw us as he walked in and sat down; he even nodded in our direction. It was a tiny gesture, nothing wrong with it. But he didn't leave it at that. As preachers preached and singers sang and a city prayed, Bush turned around from time to time to shoot us little smiles. He scrunched up his forehead, as if to ask us silently what we were up to back there. He wiggled his eyebrows, a wacky and wordless hello. These were his usual merry tics, but this was a discordant setting for them, and it was astonishing that he wasn't more concerned that one of the television or still cameras might catch him mid-twinkle.

At one point, when someone near our seats dropped a case of plastic water bottles and caused a clatter, Bush glanced back at us with a teasing, are-you-guys-behaving-yourselves expression, and he kept his amused face pivoted in our direction for an awfully long time. About twenty minutes later, he was at it again. The Rev. Al Meredith, the pastor of the Wedgwood Baptist Church, asked if everyone in the audience wanted "to see the spirit of the living God sweep over this land like a wildfire." Meredith called for raised hands, and he added, "Media, put your notepads down if you're in with us on this." Zoom—Bush was looking in our direction, eyebrows up, head cocked, the possibility of laughter on his lips.

I was taken aback, but I was not really surprised. From the time I began covering Bush in late August, my first and strongest impressions were of a man chafing against and throwing off the formal constraints of the part he had signed up for, an irreverent rapscallion on intermittently good behavior, Jim Carrey trying to incorporate at least a few elements of Jimmy Stewart. Sometimes he succeeded. Sometimes he didn't. He was asking to be king, but he still reserved the right to play the fool.

It was a role he had inhabited for much of his life, partly because it was thrust on him early, partly because he had discovered over time that it was a way to distinguish and carve a niche for himself in worlds where his other abilities were not always superior. According to some friends, the death of his younger sister, Robin, when she was three years old and he was seven had left him, for a while, not just as an only child but as one of the principal sources of consolation for his mother, Barbara. So he spread good cheer and sowed laughs, and that became his way as he moved through the stages of his youth. He was unexceptional at competitive sports and unexceptional at academics; his father's excellence in both endeavors was something he couldn't match. But he refused to be weighed down by his limitations and found an alternate path to prominence and popularity. He worked his personality, developing a reputation as a good-time fellow and dauntless prankster.

He never once made the honor roll at Andover, although 110 other boys in his class did. He stood out by cracking people up. During his senior year, he put on a top hat and rose to his feet at a weekly school assembly to announce that he was forming a stickball league—a rebellious digression from the school's hyperserious athletic traditions—and appointing himself its high commissioner. He named one team the Nads, which predictably led to the testicular game-day exhortation: "GoNads!" Another team was called the Beavers. At Yale, he impressed classmates not with his brains or his brawn but with his bonhomie. He scanned the student register at night to memorize names, and may have known more of them than any of his peers did. Perhaps inevitably, he was elected president of his fraternity.

The line between effervescence and inanity sometimes escaped him, a fact unchanged by his accumulation of experiences and his entry into politics. As Texas governor, when he held a news conference to address the subject of a fatal heat wave and related forest fires that were ravaging a part of the state, he summoned a Forestry Service official to the microphone by saying, "Tree Man, get up here!" While the official talked, Bush looked toward the journalists, stuck out his tongue and made a funny face by puffing up his cheeks like a blowfish. He gave a special wink to a *Newsweek* reporter who had come to Texas, like so many other journalists, to assess a politician who was generating and attracting considerable presidential buzz. This, apparently, was Bush's way of making a good first impression.

Even as he officially began his presidential campaign midway through 1999, he let his yen for clowning overwhelm good judgment—and, in one case, good taste. Doing an interview for *Talk* magazine with Tucker Carlson, Bush mocked Karla Faye Tucker, a pickax murderess who had been put to death by lethal injection in Texas in February of 1998. She was the first woman to be executed in Texas since the Civil War and only the second in the United States since the Supreme Court cleared the way for the reinstatement of the death penalty in 1976, and these facts, combined with her stated conversion to devout Christianity during her time in prison, meant that before her death she received widespread news media coverage. She used it to beg Bush for mercy. Bush mocked this as he kidded around with Carlson, pretending to be the condemned woman and simpering, "Please, don't kill me." Bush's aides later said that Carlson had taken encounters with Bush that were off the record and used the material anyway, and several of Bush's senior aides coined a nasty nickname for Carlson. "Mothertucker," they called him, a reference not only to how they felt about him but also to quotes in the article, which they disputed, that portrayed Bush as particularly foul-mouthed. (I myself never found—or heard—this to be true.)

There were times, early in the campaign, when Bush could be seen doing physical and emotional battle with his riskier impulses. After a news conference in Myrtle Beach, South Carolina, in late August of 1999, he walked away from the lectern and approached several of us to offer more casual greetings. He gravitated toward two familiar faces, those of Judy Keen of *USA Today* and Sam Attlesey of the *Dallas Morning News,* and kidded that he had been seeing them together quite frequently of late.

Keen made some joke that their travel itineraries were so identical they had simply started sharing the same hotel room. Bush's eyes lit up and his lips started to curl into a mischievous smile; it looked like a dozen different retorts, many of them lewd and sarcastic, were racing through his head. But then he snapped to propriety and made an exaggerated show of putting a lid on his reflexes. He flattened his lips, stared blankly ahead and, in a deliberately robotic tone of voice, said, "I'm not supposed to react to that." The routine was a miniature pantomime of frustration and determination, a commentary on the disparity between his basic instincts and his best interests.

Not too long after that, a similar performance took place outside the governor's mansion one night, when another small group of reporters mingled with Bush and his wife, Laura, during what was essentially a prolonged photo opportunity. One of us asked him a relatively innocuous question—I can't remember what—and his response, which transcended the particular situation, was to tighten his posture, clamp his legs together, straighten his arms against his sides and stand like a red-coated guard outside Buckingham Palace. Through an almost shut mouth, he mumbled, "I'm not saying nothin' about anythin'." It was a burlesque and caricature of grim seriousness.

More often, he sought to strike a kind of balance, to elevate his behavior and bearing without jettisoning the rest of it, because it was the rest of it that had drawn so many people to his side and made them like him so much. He was, to many of the political hands around him, a blast of bracing fresh air, someone who unabashedly groused about the "B.S." meeting that he had to take or the "B.S." news conference that he had to hold. He demonstrated disdain for certain political rituals, surreptitiously rolling his eyes as he emerged from pro forma sessions with the kinds of political types he was obliged to stroke and pacify. He made fun of it all and had fun with it all, and a part of him resisted any pull in the opposite direction.

He stood funny, his shoulders hunched forward, his arms curled toward his pants pockets, his hands slightly open, like a gunslinger determined to have the fastest draw. He talked funny, in elliptical colloquialisms and joking asides that weren't even in the same hemisphere as eloquence. He moved funny and approached people in a funny way, hurtling toward them, stopping short, flashing a mock-peeved expression, like he'd just found someone out, then replacing it with a smile and cracking up. He'd playact the roles of a concerned schoolmarm, a bemused co-conspirator and an adoring confidante, all in a sequence of five seconds, all in the interest of having an effect—any effect. He made his presence known and he made his presence frivolous, because he was often as interested in entertaining as in dignifying himself.

To some extent, it was a behavior wholly understandable and perhaps even explainable in terms of the eccentric and insipid rituals of American politics and, in particular, American political campaigns, which had developed into vacuous road shows filled with ludicrous conceits. I had followed a campaign only once before, and I had sworn at the time that I would never do it again. In 1996, when Bob Dole was fighting a ridiculously uphill battle against Bill Clinton, I was assigned by the *Times* to cover the first three weeks of Jack Kemp's efforts on the trail. You may or may not remember Kemp. He was Dole's vice-presidential pick, intended to add a full head of hair, a less-wizened visage and a certain political va-va-voom to the Republican ticket. The effect was catalytic, and by 1998, when the Republican horses were lining up for the 2000 derby, Kemp was already glue.

Trailing after him was breathlessly exciting. After a few joint appearances on the stump with Dole—many of them set up by black musicians performing a version of "Soul Man" that had been renamed "Dole Man"—Kemp retreated to a relative's house in Newport Beach, California. This was where he waited out the Democratic National Convention, during which the Republican ticket was guaranteed about as much attention from the news media as a Massapequa dinner-theater production starring Sally Struthers might receive. The small pack of reporters assigned to Kemp set up camp at a nearby resort hotel, where we jogged, played tennis, swam and ate long, avocado-laden lunches. Pulitzers were surely in the offing.

We kept the requisite "body watch" on Kemp, which meant that we took turns sitting in a van parked by the curb outside his relative's house. This vigil made dubious sense. The idea was that if Kemp suffered a nasty trip-and-fall or spontaneously blurted even the most oblique dissent from the Dole orthodoxy, we would be in a position to witness this critical development and relay the attendant details to our colleagues. Of course, the probability of anything like a coronary or a testimonial to the flat tax occurring right there, in the front yard, within view of the van, was fairly remote. And if it happened anywhere else, information about it would be best gleaned by telephone, which meant that we would be better off back in our hotel rooms, with reliable land lines. Still we sat by the curb, air-conditioned and inert.

When we were on the road, we were yanked hither and yon, from a rally underneath the arch in St. Louis to a rally on the plains in Montana where people sat on bales of hay. The scenery changed, but Kemp didn't, offering the same predigested, market-researched phrases in inner-city Los Angeles as he did on an island suburb in Puget Sound. A fellow reporter on the Kemp plane, marveling that I had never done this sort of assignment, said that it must seem foreign to me. Actually, I said, it did not. I explained that I had once written about movies and that a political campaign was like a vagabond movie junket that sometimes spanned three time zones in one day and offered much less in the way of commemorative graft. (I still listen to the salsa numbers on the compact disc that was my reward for interviewing Marisa Tomei before the release of *The Perez Family*.) In both cases, a product—the star, the candidate—was sold through assiduous stage managing, carefully parceled exposure and the efforts of loyal handlers who led reporters on invisible leashes and doled out frequent culinary treats. The effect in each case was equally spurious—and equally fattening.

I remember little of what Kemp said, but I remember that before and after every speech, as he sauntered toward or away from the microphone, he would lift his arm, bend it back and then shoot it forward, a gesture meant to mimic the throwing of a football. Sometimes, he would point to someone in the crowd before doing this, as if he were singling out the receiver whom he wanted to catch his imaginary pass. This was supposed to remind the crowd that he had once played quarterback for the Buffalo Bills, and that, in turn, was supposed to bolster his credentials and burnish his appeal. Perhaps his mistake was not directing the receiver to go deep. From the start, what he and Dole needed was a "Hail Mary."

If the whole process looked this silly from a reporter's vantage point, it probably looked equally so from a candidate's, and that was clearly one reason that so much of what Bush said and did came with a little thought balloon over his head: "Do we have to take all of this so seriously?" I always got the sense that his antics were in part an acknowledgment or assertion that a well-adjusted person could not approach all of the obligatory appearances, grandiose pageantry and forced gallantry toward the news media with a totally straight face. It made him likable. It made him real. Above all, it made sense: in trying to take what could be stuffy, strained situations and render them jollier and jazzier, Bush was molding them to his own demeanor, making them more palatable to him. But how well and how dependably could he mitigate that impulse? Or integrate it into something more substantial? Or, when the occasion demanded, simply let it go?

Over the course of his life, he had tamed his wild, rebellious streak, settling into marriage and fatherhood and quitting drinking at the age of forty. He had tamed the once-famous temper to which he had subjected the journalists who covered his father's 1988 presidential campaign, when George W. Bush gave a few of them memorable tongue-lashings for anything untoward, unsavory or merely unwelcome that they wrote or said. But he was less willing, or able, to tame his goofball, too-blithe spirit. And the question of whether he would make it to the White House, as well as how successful he could be there if he did, seemed to hinge on this final act of metamorphosis—on at least a partial break with this stubbornly tenacious vestige of who he had long been.

The nicknames that he doled out to almost everyone around him were merely the most widely reported example of its persistence, but they were not the best one. Much less remarked upon—and much more jarring—was the tactile element of Bush's interpersonal style, another possible holdover from his fraternity days, another facet of a fraternity-house sensibility that had never wholly abated. He touched those of us around him a lot. (I should hasten to add that the "us" refers to male reporters; Bush was visibly—and understandably—more restrained with female reporters.) He pinched our cheeks or gently slapped them, in an almost grandmotherly, aren't-you-adorable way. At least twice, on the campaign plane, I felt someone's hands closing tight on my throat and turned around to see the outstretched arms of the future president of the United States, a devilish and delighted gleam in his eyes. He once even put his index fingers in my ears to illustrate that a comment he was about to make would be off the record. On another occasion, he grabbed the sides of my head with his hands, pressed his forehead against mine and made a sound not unlike that of a moderately exasperated pooch.

Bald reporters received a different gesture, one that Bush thought funny enough that he repeated it over and over. He would lay his hands on their gleaming pates, assume an expression of ministerial concern and bellow, "Heal!" Bush's was a sense of humor that tilted toward the Austin Powers school, not just figuratively but literally. He was, for example, constantly lifting his pinkie to the corner of his mouth to mimic the Dr. Evil character in the Powers movies, which he loved, and he laughed loudly one day when a supporter gave him a photo illustration depicting him as his father's "Mini-Me."

He also loved to get in his defiant digs. Toward the end of one news conference in September, when someone asked him a question that opened the window to a mention of violence in East Timor, he paused before and after his reference to the people who lived there, underlining it for all who were listening. Just a few months earlier, the *Times* columnist Maureen Dowd had taken him to task for calling those people "East Timorians," when the correct phrase was "East Timorese," and now he was going to get it right and signal how silly he thought the attention to his flub had been.

"East Timorese," he carefully enunciated, and he searched the scrum of reporters and locked eyes with me, the only person from the *Times* on hand, as he did so. His small, pointed smile was like a challenge or taunt. Okay, he seemed to be saying, I passed the test this time. Big deal. Happy now?

The following month, after the *Times* published an article in which I suggested that he was no great reader of books, he spotted me in a parking lot, brought the sport-utility vehicle in which he was riding to a halt and lowered the window. He shouted a question about how I was.

"Tired," I said, going on to note that the first event on his schedule that day had begun around eight A.M., an hour unfit for a night owl like me.

"I got up early," Bush said, "because I was in the middle of a really good book."

He punched these last words, just to make certain I got the point, and then, reflecting on his own remark, added, "Touché!" Up went both his eyebrows and his window, and off he drove.

This was the way he handled the ever-present questions about his intelligence and degree of preparation for the presidency—by ribbing others and by ribbing himself, an approach that carried the risk of seeming flip but that he and his advisers had obviously gambled on coming across as endearing. Sometimes, as in my encounter with him in the parking lot, those jibes disproved the notion that he was not so very quick—they were clever and pointed and had just the right touch. Sometimes they fell flatter, which was the case when he visited an elementary school in Bedford, New Hampshire, in early September. He told the students: "Some people say that I proved that if you get a C average, you can end up being successful in life." More than a few teachers and parents undoubtedly cringed inwardly at this statement: it did not square with their constant nudging of students to apply themselves. But the comment provoked hearty laughter, which was Bush's favorite sound.

Schools were high on his list of must-go places that September, as they would be the next month and had been the previous month. This, after all, was an era when it was no longer enough for a politician to kiss little kids. He had to read to them and take ridiculous questions from them and watch their painfully clumsy skits and somehow mention his pledge to double their value to their parents as tax credits without making them sound like more profitable investments than tech stocks. Bush, as well as Gore, kept circling back to school gymnasiums and auditoriums and classrooms and libraries, yielding a sequence of events in the late summer of 1999 that said almost everything that needed saying about who these two men were and how they presented themselves. It was, in a nutshell, the whole story of the 2000 campaign.

In late August—indeed, just an hour or two after that news conference in Myrtle Beach—Bush appeared at an elementary school in Florence, South Carolina, and opened himself to questions from students. This did not exactly fall into the category of throwing caution to the wind, or even to the faintest of breezes. The whole point of such an exercise was to generate some precious (in more than one sense) imagery for local television stations without entering into the kind of substantive discussion that might be parsed and deconstructed for a meaning unintended by the campaign. What, after all, were the chances that a first- or second-grader might rattle or stump Bush?

Pretty good, as it turned out. And the mind-bender that did it was this: what was his favorite book as a child?

In fairness, I would not be able to answer that question reflexively, and I'm nearly twenty years closer to childhood than Bush. But I probably could, after a couple of seconds, stammer a safe, serviceable title: *Where the Wild Things Are, Green Eggs and Ham,* something with a Horton or a Hoo or a cap-clad cat. Bush went blank. He went blank despite a campaign schedule so crazy-crowded with classrooms that he was *bound* to get a question precisely like this one. He went blank despite having both a mother and a wife who had made literacy their signature issue and despite his own emphasis on the importance of education, which upped the embarrassment ante on a moment like this one and increased the importance of avoiding it. He would have been fine, with just a little foresight and preparation. But Bush was sometimes spotty on the former and cavalier about the latter.

Enter Gore, sensing an opportunity for one-upmanship that grew riper and juicier just a few weeks after Bush's bout of literary amnesia, when Bush skipped a long-planned event at an elementary school in Providence, Rhode Island, leaving scores of students in the lurch. Bad weather in the Northeast had grounded Bush's plane, thrown his schedule out of whack and put him several hours behind. He simply couldn't get to the school in time. But he did make it to the $1,000-a-head Republican fund-raiser that came next on his itinerary. And Gore—well, he couldn't resist a target that big.

Hastily rearranging his own schedule, he managed to meet the same 160 students who had fruitlessly awaited Bush's arrival just two days earlier. "Now, you're getting a lot of attention at this school," Gore told the children. "And you know why? Because your education is the most important thing to the people who came here." This list of people included the Republican governor of the state, Lincoln Almond, who had visited the school the day after Bush's no-show as a sort of consolation prize. This list of people included Gore. This list of people did not include Bush.

Of course it would be wrong to think that Gore was brandishing and then twisting a political dagger; politicians are famously above such acts of random pettiness. And Gore's response to a question from one of the children about his favorite aspect of school was *indisputably* unstudied—wholly disconnected from anything Bush had said or done over the previous weeks.

"I loved to read books," Gore said. "One of my favorite books was *Mr. Popper's Penguins.*" There it was, unsolicited and unnecessary, a precise recollection on precisely the point that had eluded Bush. You could practically see the gold star emblazoned on Gore's forehead.

Bush had little use for him. Publicly, he never said so, but privately, when he talked to friends and political allies, he made it clear that he saw Gore as equal parts pompous blowhard and preening chameleon, a spineless panderer ready to be anything for anyone. For Bush, this was distilled in a single, oddly chosen detail: "The man dyes his hair."

"What does that *tell* you about him?" Bush would ask, a question that was entirely rhetorical. Because then Bush would answer it: "He doesn't know *who he is.*" Dyes hair; has no soul: it fell short of syllogistic logic, except for Bush, who was so profoundly obsessed by it—and made this observation so frequently, and to so many acquaintances—that he was bound to let his guard down and make it publicly. And so he eventually did, noting during an appearance on *Fox News Sunday* in January that he, unlike some other candidates he could mention, didn't wear earth tones and get alpha-male lessons and *dye his hair.*

It seemed not to matter to Bush that he had no proof of this follicular fraudulence or that Ronald Reagan had, by all appearances and reports, washed that gray right out of his own shellacked coiffure. Someone once pointed this out to Bush; he did not respond and changed the subject. As Bush saw it, Gore lacked authenticity, and the proof was smack dab above his rising forehead.

Despite Bush's distaste for Gore, he liked to paint the vice president as a formidable adversary and talk about what a tough row he himself had to hoe, campaigning against an incumbent vice president at a time of unusual peace and prosperity. But this oft-repeated comment ignored his own enormous advantages, advantages that hit me full-force at the very first Bush event that I attended, a Republican luncheon in a Hilton in North Charleston, South Carolina, in late August. Racial breakdown: 98 percent white, 2 percent kitchen help. Median age: 50. Median income: impossible to say for certain, but $100,000 was probably an exceedingly safe guess.

I stopped and asked at least two dozen people why they were there; why they had chosen to support Bush; why—with the first Republican primary still six months away—they had already made up their minds and cast their lot with this relatively untested man, whose career in public office was at this point less than five years long. The two most common answers that I heard that day—and that I continued to hear for some time to come—were these: they respected his parents and, because of the money he was raising and the Republican endorsements he was accruing, they thought he could win. Sure, he was pressing other popular buttons with the Republican faithful by promising a tax cut, alluding to disintegrating social mores and—most rousing of all to these crowds—pledging to bring "honor and integrity" or "honor and dignity" to the White House. (Translation: Bill Clinton had not.) But Bush's velocity boiled down to the four letters in his last name and to his war chest of campaign contributions, which to some extent boiled down to those same four letters.

Even two months later, after Bush had given detailed speeches on education and national defense, he was still simply an amorphous but pleasant-sounding syllable for many and maybe most

Americans. In a CBS News poll, when voters were asked to utter the first thing that came to mind at the mention of George W. Bush, 38 percent said he was the son of a former president, and 6 percent misidentified him *as* a former president. Only 5 percent mentioned that he was the governor of the nation's second most populous state. A subsequent poll in mid-November by CBS News and the *New York Times* showed that support for Bush was less firmly grounded in anything real than support for some other candidates was. While 86 percent of the respondents who said they favored Senator John McCain could cite a specific reason, only 61 percent of those in Bush's camp could do so.

It was all very odd, because it suggested the passive munificence of Poppy, as George Herbert Walker Bush was known in the family, and Poppy had left the White House after one tumultuous term with his tail between his legs. To the Republican faithful, the elder Bush was a traitor who had asked the nation to read his lips and then abruptly told them to forget what they had seen there, signing off on precisely the kind of tax hike he had promised to oppose. He had alienated conservatives. He had also engendered suspicions among Americans in the political center that he was out of touch with their concerns, and this opened the door wide for Clinton and Ross Perot. The month before the 1992 election, the inexorably ebbing president was viewed favorably by only 29 percent of the electorate, and he was viewed unfavorably by 49 percent.

But Travelgate and Whitewater and Monica Lewinsky and Ken Starr changed things, embittering and disgusting even many of those voters who thought that President Clinton was doing an otherwise good job. Nostalgia took root. So, in a sense, did reason: the first President Bush's reversal of his tax pledge had arguably been a brave if self-destructive act, the necessary prelude to the budget balancing and debt reduction that became the vogue—and revitalized the economy—over the next eight years. What many people seemed to remember about Bush One was its relative dearth of scandals, its relative surfeit of patrician warmth. The political operatives working to make Bush Two happen knew this—knew that the surname now yielded positive associations, that it alone bought their candidate instant good will. When they cut television commercials for him, they didn't bother with bits of the candidate's biography to introduce him to voters; his patrimony had laid the groundwork for him. It had, for whatever complicated reasons, established the baseline trust and respect he needed. It made him, by his advisers' own candid admission, a popular "brand," one that was coming along at just the right national moment, peddling plainness, good cheer and a deficit of cutthroat ambition and complicated passions to a country looking for a little relief.

Timing was everything, and Bush, with his buoyancy of spirit, presented a perfect contrast to Gore, so turgid and tendentious. Republican power brokers loved this about him. And they loved his departure from the brooding of a Bob Dole or the belligerence of a Newt Gingrich, a break that complemented his marketing of himself as a "different kind of Republican" and gave him a fighting chance to connect with voters in a warmer, less ideologically freighted way.

So they sent him out there, clucking over his charisma, although this was not really what he had. Charisma connotes enough palpable magnetism to wow an entire crowd at once; it suggests an electricity of bearing that undulates across a large room, sparking someone far on the other side, or that crackles through the television set. Bush's wattage was more modest and quirkier, its sphere of illumination more discreet. What he could do—with sometimes-clever banter, frequent self-effacement, playful shenanigans and an aggressive friendliness that felt enticingly like validation—was collapse the distance between his perch high on a political pedestal and the lesser altitude of the person beside or in front of him. It was a one-on-one thing, or a one-on-three thing, but not a one-on-a-hundred thing, not unless Bush was *really* on. As Ed Gillespie, a Republican political consultant who worked on the Bush campaign for the final six months of it, once said to me, "He's a much better retail politician than a wholesale politician."

A visit he made in mid-November to a Timberland factory in New Hampshire, which included a carefully choreographed tour of the premises and brief conversations with select employees and other people, was a perfect example. When he praised a twenty-year-old man for his "articulate" remarks, he appended this high-minded compliment with an unexpected term of endearment.

"Dude," Bush called his new acquaintance.

When he crossed paths with an elderly employee and she told him that he had her support, he said, "I'll seal it with a kiss!" Then he swooped down on the captive seamstress and made good on his promise.

Bush's arm curled tight around the shoulders of other voters; he arched his eyebrows and threw coquettish grins and conspiratorial glances their way. It was campaigning as facial calisthenics, and Bush was its Jack LaLanne.

This gift, or whatever it was, partly explained why Karl Rove, the lead strategist not just of the Bush presidential campaign but of the Bush political career, had serially lured small groups of important thinkers, wealthy backers, corporate chieftains, prominent Republicans and other opinion makers down to the governor's mansion in Austin, Texas, during the spring of 1999 so they could talk with Bush in a more casual vein. There was a practical consideration as well: Bush was immersed in a legislative session, and since these came along only every two years in Texas, they could not be skipped, even by someone planting the seeds of a presidential campaign. But the meetings in the mansion allowed the participants to see Bush up close and personal, which was how he was most comfortable, nimblest and, not so incidentally, most impressive.

I knew about that last part because I came to see it with my own eyes. On several of the nearly dozen occasions over the following two years when I would interview Bush, he would keep me around for a short while after his aides had told me to shut off my tape recorder and talk to me, more loosely and freely, off the record. For that reason, I can't reproduce the conversations; I have neither an accurate enough record nor the ethical license to do so. But I remember that as he talked about why he was hewing to a certain strategy or the way certain events were shaking out, he came across as instinctively bright, quick to the punch. Not only did he do better in small groups than large groups, but he did better in private situations than in public ones, and it seemed like the exact opposite of how a politician—in order to succeed—should be.

In order to succeed, for example, a politician running for national office might be expected to handle himself well on the stump. It was where he had to spend the lion's share of his time. It was the manner in which many voters would see and hear him for the first time. In Bush's case, it was also a kind of metaphor for the formal requirements of leadership, artificial but ineluctable, that he would have to try to master. For a good long while, Bush was woefully—and sometimes hilariously—unable to.

CHAPTER TWO

YOU WANT TO RUN FOR PRESIDENT? HERE'S what you need to do: Have someone write you a lovely speech that stakes out popular positions in unwavering language and less popular positions in fuzzier terms. Better yet if it bows to God and country at every turn—that's called uplift. Make it rife with optimism, a trumpet blast not just about morning in America but about a perpetual, dazzling dawn. Avoid talk of hard choices and daunting challenges; nobody wants those. Nod to people along all points of the political spectrum, because you're going to need broad support. Add a soupçon of alliteration, which rolls nicely off the tongue. Sprinkle with a few personal observations or stories—it humanizes you. Stir with enthusiasm.

Bush had this much down pat. Right near the top of his stump speech, he would mention his then-seventeen-year-old daughters, fraternal twins, and recall that one of them recently told him,

"Daddy, I've been reading about you in the newspapers. And I've been reading about those polls. You are not nearly as cool as the people think you are." (Cue the laughter.) Then, in case anyone doubted it: "I feel incredibly fortunate to be an American. I love our country. I love what our country stands for. I love Americans." (Cue the applause.) A statement of confidence: "We must seize the moment and lead the world to peace, which we can do." A poetic phrase: "America has never been proven by cities of gold, but by citizens of character." Something for everyone: an endorsement of tax cuts for fiscal conservatives, a passage about the power of churches (and synagogues and mosques) for religious conservatives of all stripes, a promise that "no child is left behind" for swing voters who sometimes see Republicans as uncaring, bereft of heart and soul.

Without uttering the word "abortion," Bush pleased pro-life voters by saying, "I believe our nation ought to usher in what I call the responsibility era, an era that will stand in stark contrast to the last few decades." He discussed giving the parents of children in failing schools money to seek private options, but never—ever—used the loaded, politically controversial definition of what this approach really was: a "voucher" program.

Mission accomplished. Now for the less agreeable—but no more intellectually challenging—obligation: be willing to repeat these remarks ad nauseum, not just ten times or a hundred times but more like a thousand times, to the point where you could be woken up in the middle of the night and spill them out. Fly from Columbia, South Carolina, to Everett, Washington, and say them, and then fly from Everett, Washington, to Fargo, North Dakota, and say them again. Learn to like those earth-toned, soulless ballrooms in Holiday Inns and Radisson hotels far and wide, because you're going to visit scores of them. Be prepared to stand on the edge of a marina in Hilton Head, South Carolina, or inside a production plant in Saginaw, Michigan, because each has a constituency you want. Reach the point where you consider lecterns and handheld microphones everyday appendages, because they will be. Think of your speech the way you think of meals: at least three a day is best. Do not lapse into a self-induced coma, and remember that political campaigns are not to be essayed by anyone who craves genuine excitement, variety and a cauldron of new ideas. The applause and the sound of your own voice will have to sustain you, because you are a broken record. Play it again, ham.

Bush did, and we listened and typed, desperately trying to stay awake, dutifully loading and unloading our laptop computers every few hours, putting new batteries and blank tapes in our recorders, teetering on skimpy folding chairs in the back rows of overheated gymnasiums and auditoriums and office-building atriums. It was journalism as aerobic stenography, a constant rush from one location to the next so that we could do yet again what we had just done: document what Bush was saying. Each new audience had not heard it all before, but we had, and this was undoubtedly one reason—and not an especially noble one—that we came to focus so much on his unintended and unfortunate torturings of the text: it caught the ear and quelled the boredom. But there was, in hindsight, a better reason to point out these flubs, although I'm not sure any of us saw it as such at the time. Bush was pursuing an endeavor for which one of the most basic, useful skills was a polished, mellifluous manner of public speaking, and he did not possess it.

Not since Yogi Berra—or maybe Dan Quayle—had a public figure produced such a bumper crop of bloopers. "Potential missile launches" became "potential menshul losses" and "tactical nu-cu-lar weapons" (he could never pronounce "nuclear" correctly) morphed into "tacular nucular weapons." He sympathized with the difficulties that some Americans faced in trying "to put food on your family." He feared that excessive regulation of education from Washington put "this kind of federal—federal cufflink" on local schools and teachers. He also believed that the country should pursue free-trade policies that knocked down not just tariffs and barriers but also "bariffs and terriers," and the spaniels and retrievers could only wonder if they would be next.

The on-line magazine *Slate* began a running tote board—later compiled by Jacob Weisberg into a book—of Bush's latest and greatest misses. An ABC producer's daily written report to the network's news division included a regular feature titled "The English Patient." It was all captured here or somewhere else, each trip of the lip, every grammatical bobble: the time when he called Greeks "Grecians" and the people in Kosovo "Kosovians." When he referred to a "vile" instead of a "vital" hemisphere and to tailpipe "admissions." When he insisted that quotas would "vulcanize" instead of "Balkanize" society. When he described his approach to the economy as an effort to "make the pie higher," a theoretically sensible strategy if he was thinking in terms of meringues or soufflés. When he said that the salient question behind any education policy was, "Is our children learning?"

Some errors were more sustained and embarrassing than others. Extolling character education during a January visit to an elementary school in New Hampshire that was celebrating "Perseverance Month," Bush kept saying "preservation" instead, making the observation that a presidential candidate quickly learns: "You've got to preserve." Perhaps, but only if he plans to take homemade jams and jellies with him on the road.

It eventually reached a point, much later in the campaign, when the journalist-cum-pop-psychologist Gail Sheehy, writing in

Vanity Fair, would postulate that Bush suffered from dyslexia, a diagnosis that one of his brothers, Neil, actually had. There were problems with this hypothesis, chiefly that Bush seldom made terrible gaffes when reading from a TelePrompTer; he seemed to see and enunciate the words just fine. But the publication of the article and the questions he briefly fielded about it led to several of the campaign's most surreally funny episodes. Denying Sheehy's quasi-allegation during an impromptu news conference that also strayed onto other topics, Bush repeatedly mispronounced the word "subliminal," inserting an extra syllable into it and saying "subliminable." Days later, he told some voters he tripped across, "The woman alleged that I had—said I had dyslexia. I never interviewed her." You couldn't make this stuff up.

The presidential campaign of 2000 was not a laboratory for impressive oratory. Al Gore tended to drone on and on, in singsong, narcotizing cadences best endured by the heavily caffeinated. Steve Forbes came courtesy of a political animatronics laboratory, mechanized and monotonous. And Bill Bradley? Little of the verve and agility he had once exhibited on the basketball court carried over to the campaign trail, where he shuffled along like a rumpled, distracted professor, the kind whose classes were taken only by students who needed to fulfill the requirements of their major. Clinton nostalgia was inevitable: none of these guys filled a room the way he did. None of them gave it his kind of spark.

But Bush wasn't just flawed when it came to delivering previously written, endlessly repeated remarks; he was—in public although not in private—sometimes miserable at spontaneity. Campaigning for president required this, too; you could have your predigested comments and recite them from memory, but every so often, you had to let your audience ask you a few questions. And when Bush did this, his instinct was to pluck anew from his stump speech, to retrieve a few lines that might somehow apply. If that didn't work, he occasionally segued into diffuse generalities, and in some cases, pure babble. Both were on display one day in early November when he opened the floor to the gathered citizenry of Gorham, New Hampshire, who had just heard him deliver an address about character education in elementary schools, stricter discipline and related matters.

Someone asked him about "his feelings on the new economy and how rural communities can invest their hopes in the future through the information superhighway." What followed was an oratorical journey so amorphous and interminable that it received scant attention in news reports, including my own: it was just too time-consuming, too outrageously beside the immediate point.

Bush started by saying that "we're in an incredibly exciting time" and continued: "The whole nature of the economy is changing in ways that are hard to envision. That creates a real dilemma for Americans. That's why I spend so much time talking about education. The great challenge for America is to make sure the American dream touches every willing heart." This last bit wasn't some nifty, just-hatched phrase but a veritable mantra that he had been using for months, and he kept circling back to it as if it were a life raft to which he was clinging for support.

"As the nature of our economy changes," Bush went on, "if our children are not educated not just in reading and writing but right and wrong, the American dream will not touch every willing heart. The big challenge for America domestically is to make sure people are literate enough to access where the economy is headed. Internationally, the nature of the new economy is going to create all sorts of interesting opportunities and problems. The interesting opportunities are—capital will move freely when we're a global nation, in a global world. We're a nation in a global world." Was there some other kind of world, perhaps shaped like a pyramid? What did the movement of capital have to do with the Internet?

Bush wasn't saying. "The ability to communicate and capital to move quickly because of the new economy is changing the nature of the world," he continued. "People are going to resist the flows of capital the likes of which we've never seen before, which is going to create tension—will create a sense of uncertainty on the one hand, but uncertainty on the other." Those two hands looked mightily alike, and neither seemed to be placed on the keyboard of a personal computer in Gorham.

At long last, he got to the Internet, albeit with no real segue—it was as if he snapped out of a global trance and finally remembered the question. "I believe that rural Texas and rural New Hampshire are going to benefit mightily from the new economy once the highways are laid," he said. "The reason I believe that is, people oftentimes really are asking the question, 'Where do I want to raise my kids? Where's a community in which I can find neighbor loving neighbor? Where's a place where if somebody sees my child running across the street, they won't hesitate to say: look both ways before you run?' Oftentimes, those values are found in rural America."

He was getting somewhere, but the voyage was herky-jerky and much delayed. "The ability for people to find work is going to change when the Internet makes it to your community," he explained. "I believe education will be changed dramatically by the Internet. I mean, we're going to be able to bring information into the classroom the likes of which you've never seen before." Back to a stock phrase, another flotation device.

"I thought you were going to ask me about home schooling,"

Bush said, apropos of nothing. He then dwelled briefly on the Internet's implications for home schooling, including his belief that "art will be able to be brought to—to—the community the likes of which has never been seen before. Libraries will change."

He created a geographical conundrum: "Health will change. In Lubbock, Texas, the other day, a relatively big West Texas city, I saw a man in Alpine, Texas, having his ear examined through the Internet. My only point to you is that when health and education and job opportunities can be found over the Internet, it's going to change the quality of life, it's going to change the opportunities available. Children graduating from high school are more likely to find exciting, high-tech, high-paying, high-quality jobs closer to home and won't go to big cities, so I've got great hope."

He wasn't finished—not even close. "In terms of the highways, we've got what's called telecommunications infrastructure within our state. And I understand, by the way, that markets are more likely to first lay capital where there's a density of population. The—the—the technology, by the way, is changing the disbursal of the highways." He added that satellite-type "dishes will be able to carry enough bandwidth to have two-way communications, hopefully, someday, which will be able to solve part of your problem, whether or not you can get bandwidth through your telephones and/or cable TVs." Satellites? Bandwidth? Austin, we have a problem.

"Anyway," he explained, "my only point is, one of the main reasons I'm running for president, though, is to remember that as the nature of our economy changes, we better be careful because people could be left behind. And I thought you were eloquent—not to be overly pandering—but I thought you were eloquent when you were talking about the American dream. It is the power—it is the power of our nation, it is the strength of our country—the American dream can touch every willing heart." And so he ended as he had begun, with that ever-groping dream and that ever-pliant heart, his perennial oratorical valentine.

Deep within this bramble of ramble, Bush answered the question, and did so with some fundamentally correct, fleetingly cogent observations. But there was all of that other pap, the flailing of a nervous, befuddled student who had not expected this particular test and had not pre-formulated a response he could give at the ready. And so he larded his answer with irrelevant minutiae and trite platitudes and swam furiously toward a familiar shore, desperate to get back to the chosen topics of the moment, the ones he had been prepared to address: education, better opportunities for all Americans. And he drowned.

It was an extreme example, but a telling one. Cautious to the point of timidity, disciplined to the point of paralysis, Bush wanted to say only what he had decided to say; he did not like to improvise and ventured into such waters at his own peril. Two months later, in the best recurrence of a Gorhamesque moment, he tried to inject a few fresh notes into a stock melody about the changed world order, about the vanished Cold War paradigm of the United States versus the Soviet Union. The result was a marvel of free-floating pronouns and absent antecedents.

"When I was coming up," he told a crowd in Council Bluffs, Iowa, "it was a dangerous world and we knew exactly who the 'they' were. It was us versus them and it was clear who 'them' was. Today, we're not so sure who the 'they' are, but we know they're there." I glanced at the other reporters present to see if their astonished expressions matched my own and noticed that Attlesey of the *Dallas Morning News*, a strapping, chain-smoking veteran of political reporting, actually had tears of laughter streaming down his face. We all huddled to make sure we had heard and recorded Bush's words just right, and Karen Hughes, the campaign's communications director, hustled over to find out what the fuss was about, peering over some of our shoulders to see what we were so busily punching into our laptop computers. Her expression somehow remained calm as she opined that no matter how Bush's words had emerged, people would understand what he was saying.

On his good days, Bush did well enough, and on his best days, he did better than that, shocking and seriously impressing us. I was sitting with fellow campaign reporters in a college auditorium in another New Hampshire burg when a student asked Bush about the mapping of the human genome, a topic he had not publicly addressed before. I could see my colleagues widen their eyes and draw in their breath, steeling themselves for catastrophe, and I quickly shot a glance at my recorder to make sure it was working. This promised to be priceless. Would the genome be a vehicle to make sure the American dream was touching every willing heart? Alas, no. Bush knew what it was, and said that the ability to see into people's medical futures would present excruciating ethical dilemmas and tough decisions about health insurance. Bingo.

Moments like these made it harder to get a real handle on him, on what he might be capable of and what he most certainly was not. They were also not that common, and the explanation that he gave for his lack of daring in his public comments was that his job was not to dazzle. It was, as he saw it, to iterate and reiterate his priorities and convictions, without falling prey to the temptation for variation, until people began to hear, understand, believe and agree with him. (He made it sound like a kind of brainwashing.) But was this the full reason, or was Bush insulating himself from danger, developing a rationale that justified both his shortcomings and a failure of nerve? His steadfast adherence to pre-fabricated locutions could be read as a commitment to mediocrity, a phobia of risk. In fact, many of the themes, thoughts and phrases in his presidential stump speech could be traced all the way back to 1996 or 1997, the middle of his first term as Texas governor.

He had talked then about replacing the ethic of "if it feels good, do it" with "the responsibility era," and he was saying the same thing, in the same way, in the summer and fall of 1999. He had touted the "armies of compassion" that exist in churches and synagogues, and he was still doing it. He had appended to his exhortation that young, unmarried people not become parents the promise—made twice in a row, for emphasis—that the babies would nonetheless be loved in Texas. Now, he was appending to the same exhortation the promise—again made twice in a row, for emphasis—that the babies would nonetheless be loved in America.

"We'll love the babies," he would say. "We'll love the babies." He was so faithful to this repetition that on the few occasions when he loved the babies only once, even a few of his aides would drop their jaws and join reporters in wondering if Bush had just run across a toddler who irked him. He was holding back on the love.

In treating this as a comical matter, those of us covering him may well have given short shrift to some of the questions it raised. And these questions were not so much about his intellectual agility, for which there are few reliable yardsticks, but about his core confidence in himself, his engagement in a broad arena of knowledge, his willingness to dig for something deeper and better. He often seemed content to get by on as little as possible, and we perhaps focused less on this than we might have in the fall of 1999 because his failings on the stump didn't fit the narrative in place at the time.

News coverage thrives on coherence, black-and-white truths

and tidy plot lines, and when one emerges, few reporters want to find themselves on a different, lonelier page. If a politician was rising, everything he or she was doing must be right, so we wrote not just about the record setting financial donations and the surname that were propelling Bush; we wrote about what a deft campaigner he was, how natural and loose. He was natural, but this often meant that he was naturally unremarkable. He was loose only in manner and controlled in most other ways, but this only drew sustained attention—and was suddenly framed as the great weight dragging down his campaign—when McCain surged in the polls in New Hampshire and Bush's numbers there started looking shaky. All of a sudden, *this* phenomenon cried out for explanation and presented the possibility of an even better story, one with actual suspense: that Bush's grip on victory was slipping, and that it was slipping because his style of campaigning was flawed and riddled with newly apparent vulnerabilities.

It was all partly true and partly imagined, motivated by our desire to create and live in a political universe more interesting than the one that really existed. Bush's style was what it was, and changed much less over time than the up-again, down-again, up-again media appraisals made it sound. McCain's prospects were what they were, and the head of steam he was building in New Hampshire did not change the dynamics that made it pretty much impossible for him to wrest the Republican nomination from Bush.

After McCain won this state, he graced the covers of the three major news magazines: *Time, Newsweek* and *U.S. News & World Report.* All but lost in this journalistic orgy were a plethora of cold, hard realities: that McCain, unlike Bush, had largely ignored the rest of the primary states and taken temporary residence in New Hampshire, while Bush continued to play on a broader field. That McCain had tenuous ties to religious conservatives, so powerful in Republican primaries in the southern states. That Pat Buchanan had won New Hampshire in 1996 and no one ever thought he was shaping up to be the party's nominee. That almost the entire Republican machinery had clanked into place behind Bush, and it was prepared to crush McCain, as it eventually did. This was a lesser story that involved set forces and outcomes, so it was told in the margins, in the "but" and "to be sure" paragraphs that were tucked dutifully into the middle of articles and newscasts that could then return to more enticing fare.

There was, moreover, a case to be made that Bush was actually benefiting from the rigorous sameness of his remarks over time—from what his aides and Republican political analysts liked to call his "message discipline," a euphemism if ever there was one. For all his advisers' complaints about the sins of the news media, most mainstream journalists felt an obligation, first and foremost, to report what a candidate was saying before they went on to parse, analyze and belittle it. So Bush tried to make sure that he said only so much; he made journalists feast on a main course of his most carefully considered sentences and most carefully constructed themes by not putting anything else on their plates. His aides were similarly loath to provide side dishes or garnishes. If you asked them what Bush had for breakfast, the reply might be, "The governor believes that no child should be left behind," or, "The governor wants the American dream to"—you guessed it—"touch every willing heart." Terrific, but what was it—eggs Florentine or Grape Nuts? "The governor," came the response, "believes that the country must have tort reform."

I exaggerate, but less than you might think, and it was a measure of reporters' exhaustion and frustration that whenever Bush deviated from a rote recitation of all-too-familiar words, we searched vigorously for profound meaning in the change. For several days in Iowa, Bush altered the usual conclusion of his speech, a bit about taking the oath of office and pledging to uphold not only the laws of the land, but also "the honor and dignity of the office to which I have been elected, so help me God." Suddenly, God was dead, and "elected" was the end of it. It happened once, then twice, and none of his aides could explain why, although we pestered them for an answer. Was Bush responding to some pundits' commentary that the presidential candidates were trafficking excessively in religion? Why would he do so in Iowa, where Forbes was flaunting his new anti-abortion credentials in a bid for support from the many religious conservatives who voted in the state's late-January caucuses?

John Berman, the ABC producer who wrote the installments of "The English Patient" for his colleagues at the network, was most acutely rattled, and when he next came within earshot of Bush aboard the campaign plane, he demanded satisfaction. Bush seemed amused, dismissing the mystery as absurd. He said he had simply shuffled the deck for the sake of shuffling it, and he promised to restore the Almighty to his rightful throne. He indeed did so, in the very next speech he gave, pausing before the climactic phrase, surveying the audience to find Berman, locking eyes with him and then letting it rip: "so help me God." Bush had returned to the Holy Land, and he stayed there for most of the rest of the campaign.

For several months, he had held brief news conferences on an almost daily basis. But this came to an abrupt halt after an unpleasant session in Iowa. Beforehand, those of us in the press corps had decided not to flit from one topic to the next but to try, with question after question, to get Bush to explain whether the "strict constructionists" he constantly vowed to appoint to the Supreme Court would likely interpret the Constitution in a manner that supported abortion rights or undercut them. Bush bobbed, weaved, seethed and never really answered the question, and for the next several days, there were no more news conferences.

In a candid moment, one of his spokespeople, Mindy Tucker, told a group of reporters on a campaign bus that the campaign had determined that a daily news conference was no longer "in our best interests."

"We have a message of the day and we're going to stick to it," Tucker said. "We are not going to have one big, fat news conference on our schedule where you can come and ask what you think is the news of the day."

We started scribbling, and Tucker grew frantic with worry, and soon enough her boss, Hughes, was mingling among us to try to do damage control. Among ourselves, we dubbed the incident "Mindygate" and laughed about it, but it was a tacit admission by the Bush campaign that Bush was ill-equipped to deal with the unpredictable, that it frightened and rattled him.

It sometimes felt as if he were being dragged, kicking and screaming, toward exercises that were an inescapable element of campaigning, an intrinsic part of the bargain. In the buildup to the Republican primaries, as in the buildup to the general election almost a year later, he did not seem to put much stock in debates. His offhand comments about the many scheduled Republican forums at which McCain and Forbes would get to take shots at him suggested a perspective that people who knew him well confirmed: He deemed debates to be artificial constructs that prized a cheap kind of fancy talk over any real indications of a politician's ability to lead. This thinking was perhaps inherited from his father, and perhaps stemmed from their shared struggles with eloquence—apparent, in the elder Bush's case, during a 1984 vice-presidential debate against Geraldine Ferraro, the history-making female running mate of Walter Mondale, the Democratic presidential nominee.

"Almost every place you can point, contrary to Mr. Mondale's—I gotta be careful—but contrary of how he goes around saying everything bad," then-Vice President Bush said during the forum. "If somebody sees a silver lining, he finds a big black cloud out there. Whine on, harvest moon! I mean, there's a lot going on, a lotta opportunity!" Eight years later, when then-President Bush was trying to salvage his flagging reelection campaign, he turned in a curious performance at a town hall-style encounter with Clinton and Ross Perot. The format put the candidates on stools and allowed questions from voters in the audience, and it all felt strange and contrived to Bush, who more than once stole glances, caught on television, at his wristwatch. It looked like he was willing the minute hand to spin around faster and bring the ordeal to an end, and it did little to revivify his prospects.

Former political aides said that he found the setting and the routine an irrelevant way to measure a leader's mettle; he resented being submitted to it. He telegraphed as much in interviews for a two-hour PBS special, "Debating Our Destiny: Forty Years of Presidential Debates," that aired in September of 2000, which brilliantly coincided with the resolution of a weeks-long, failed attempt by his son to dictate his own, comfort-inducing terms for his presidential debates against Gore. In response to a question in the documentary about whether debates should be "a required part of the process," the elder Bush answered: "Nope. I think you ought to do what's best to get you elected. And if that's best that you have no debates, too bad for all you debate-lovers. If there was a guy that couldn't, say, couldn't finish a sentence and yet is a brilliant contribution as a public servant or an academic or whatever, why should that one thing be mandatory? Why should a person be burdened with that decision to have to debate?"

His son seemed to agree with this assessment, and there was something arrogant about it—about applying for a job that came with certain prerequisites and wanting to dismiss those prerequisites as immaterial. He skipped the first debate of the Republican primaries, held in mid-October, citing a scheduling conflict: Laura Bush was being honored at an event at her alma mater, Southern Methodist University, on the same night. She was an only child with only one surviving parent and without an extended family like his own, and he said there was no dilemma: he simply had to be there for her. But it may not have been the open-and-shut case he made it out to be. More than a year later, at a post-election forum in Philadelphia, during which advisers of various candidates deconstructed the presidential race, Rove said that he wished he had not let Bush miss the debate. The remark suggested that the matter had not been set in stone, that Bush could have been compelled to attend if he and other senior officials in his organization had felt it was necessary.

They did not, and Bush also did not feel a compulsion to prepare for the second and third Republican debates, held in early December, as extensively as he needed to. One of his advisers told me that he was stunned and alarmed by how casually Bush approached these debates, allotting little time for full-fledged practice sessions and basically swaggering toward a pesky obligation. Several hours before his first debate, on December 2, in Manchester, New Hampshire, Bush stepped off his plane and a reporter shouted a question about what was "at stake" in that night's forum. Bush's flippant response: "I like mine medium rare."

Almost as soon as the event began, Bush's lack of preparation was obvious, as was his reserve: he was trying harder not to get anything wrong than he was to get anything especially right. There were six candidates dividing an hour and a half, meaning that no one got more than fifteen minutes in all. Even so, Bush filled up less than that; his diaphanous answers sometimes ended many seconds before they had to, creating moments of silence and awkward pauses before the moderators realized that he was done and they should move on. At every opportunity, no matter what topic he was supposed to address, he pivoted into a regurgitation of lines from his stump speeches—a tactic that all of the candidates used, but that few did as transparently and consistently as Bush. The following exchange was classic:

"Governor," asked one of the questioners, "you have said that you would always have people around you, as president, who would be able to fill in any gaps in your knowledge of the world and the players on the world's stage. Do you think that President Bush could have done the job he did in assembling and holding together the Gulf War coalition, composed of many very varied nations, had he not had the knowledge of the world that he had, from years of experience in diplomacy and politics at the UN?"

"In order to be a good president," Bush responded, "when it comes to foreign policy, it requires someone with vision, judgment and leadership. I've been the governor of the second-biggest state in the United States. If it were a nation, it would be the eleventh-largest economy in the world. I was overwhelmingly reelected because the people in my state realized I know how to lead, and I've shown good judgment. . . . My goal, should I become the president, is to keep the peace. I intend to do so by promoting free trade, which, in my judgment, promotes American values across the world. I intend to do so by strengthening alliances, which says, 'America cannot go alone.' We must be peacemakers, not peacekeepers." He went on a bit from there, but strayed no closer to the heart of the question. Finalists in the Miss America pageant had done better on the fly.

And then there was the problem with his mouth, which contorted into a weird, pursed shape that some journalists and opponents branded "the smirk." It was an expression that Bush often flashed when he was under fire, and while it was mostly a sign of discomfort, it came across as cocky, which did him no good, especially when he was simultaneously demonstrating that he had nothing to be cocky about. Bush was not the kind of actor that President Clinton had been—he would never have pulled off the Tony-worthy "that woman" soliloquy—and he was certainly not in a league with President Reagan, who had managed to look authoritative even when opining that trees caused pollution. Bush shuffled. He stammered. He smirked.

The low point came at the next debate, on December 6, in Phoenix. He had said during the Manchester forum that he was reading a biography of former secretary of state Dean Acheson, and he was asked this time around what he had learned from it. His answer made a beeline for his stump speech, from which he stole phrases that weren't necessarily inappropriate in terms of the Acheson book but that also would have been appropriate for dozens of others. He offered nothing specific, nothing original; he just stitched together sentences that he had said dozens or hundreds of times before, and if many members of the television audience weren't likely to detect this, they were sure to learn about it from the journalists who regularly tracked Bush.

But they probably did get some inkling of their own. The centerpiece of Bush's answer was that the lesson of Acheson's experience and service was that "our nation's greatest export to the world has been, is and always will be the incredible freedoms we understand in the great land called America." This was, of course, Stump Speech Central, and Bush tipped his hand just a few minutes later, by bringing it back out for a command performance when he talked about the relationship between the United States and China. "Look," Bush said, "in my earlier answer, I said our greatest export to the world has been, is and always will be the incredible freedom we understand in America." Here he was, once again sharing the stage with too many candidates given too little time, and he was using his precious portion to allude to his own previous remarks.

To some observers, the evidence was stark and irrefutable: Bush was a bozo. This judgment was in fact willfully selective and oblivious to a contradictory body of facts. The naysayers focused, for example, on Bush's college transcript, which had been leaked to the *New Yorker* and showed him to be a run-of-the-mill C student. They paid less heed to his SAT scores, which the magazine had also obtained. Bush got 566 out of 800 on the verbal portion and 640 on the math, for a respectable total—in an era before rampant grade inflation and an upward readjustment of all marks on the SATs—of over 1200. Bill Bradley, by contrast, scored below 500 on the verbal portion of the SATs, as another publication later revealed, yet journalists did not deride his admission to Princeton as a privilege of prowess on the basketball court the way they derided Bush's admission to Yale as a privilege of his pedigree. (Bush's father and grandfather went there.) As for the grades that Bush received, they were, in fact, no worse than the ones Gore got at Harvard for a long stretch of his time there, and they were better than the ones McCain got at the U.S. Naval Academy.

But the negative thread that ran through all of the coverage of Gore's character was about a runaway ambition that sometimes skirted principle and fabricated personal facts, while the negative thread (inasmuch as there was one) for McCain was about a runaway temper. For Bush, it was about buffoonery, and anything that supported it bubbled to the top. His performances at the first two debates that he had participated in—coupled with all of his misstatements and perfunctory discussions of policy—played into this impression, and Bush was being ridiculed from all sides, in ways both subtle and overt, as the guy wearing the tall, pointy dunce cap.

In the Phoenix debate, Alan Keyes, the former ambassador, spoke unspecifically about candidates who seemed merely to repeat scripted lines and might not get anywhere without their family connections. (Who, oh who, could he be thinking about?) McCain cannily found room in one of his answers to talk about Acheson in concrete, knowledgeable ways, underlining the vapidity of Bush's response. (As McCain did so, one of his spokespeople, Dan Schnur, turned to the reporters watching the event and flashed a devilish smile.) And Senator Orrin G. Hatch, a Utah Republican, said that Bush indeed had a place in Washington—as second fiddle to him. "You should have eight years with me, and, boy, you will make a heck of a president after eight years," Hatch said.

Comedians and cartoonists were more direct. Garry Trudeau's rendering of Bush in *Doonesbury* strips was an enormous cowboy hat with nothing but an asterisk beneath it. Jay Leno, referring to Bush's failure during a televised pop quiz to name lesser-known world leaders, quipped: "George Bush released his new slogan today: 'He'll Get Tough with What's-His-Name.' " David Letterman cracked, "The guy may have 'bonehead' stamped all over him."

Bush pretended not to care, implying that this caricature of him was so far from the truth it did not warrant a response. In late December, he told those of us in the press corps at a news conference in Manchester, New Hampshire, "In the '94 campaign, the same thing happened. People said he's Daddy's boy and they said he's not smart enough to be governor, he hasn't had the experience. Look who's standing here as governor of Texas."

He was more or less saying: "Go ahead, underestimate me. You'll learn. You'll be sorry." His advisers were terribly fond of this spin. One of their favorite refrains was that Bush *liked* it when people didn't realize how truly formidable he was, because then his opponents, whether in the Texas legislature or the Republican primary, never saw him coming. If you extrapolated just a bit from this rationale, Bush was underperforming just to fake people out, to make them look the other way. *Hah! Gotcha! I really DO know the names of the last two presidents of Botswana.*

The problem with this was the same as the problem with Bush's stated rationale for his ever-tight clutch on his stump speech: was the explanation for real, or was it an elaborate attempt at diversion from his failings? So much of Bush's candidacy—and, later, the beginning of his presidency—seemed to be constructed around or dictated by his weaknesses, designed to present them as something other than what they really were. We were to believe that he gave succinct public remarks not because he wasn't particularly good at longer ones or because he was trying to get through his day a little faster. He did it because he valued brevity and getting to the point. We were to believe that he paid limited attention to details and fine points not because they bored him. He did it because he never wanted to lose sight of the Big Picture, because he wanted to keep his eye on the ball itself, not on the stitches sewn into it. We were to believe that he liked to surround himself with more experienced people, delegate tasks and responsibility to them and then step out of the way not because it papered over his lack of knowledge or the limits to his energies. He did it because it was an effective management style.

All of these explanations held up to a point. But what made them seem potentially counterfeit was how ready Bush and his aides were to crow when he acted in the opposite fashion—when he flexed a familiarity with the nitty-gritty of a given policy, when he displayed some oratorical skill. Suddenly, these non-essential attributes were to be heralded and celebrated. Indeed, when he turned in a better performance in his third debate—in Des Moines, Iowa—than in his first two, his advisers' reaction was not to bemoan the way he had torn off his camouflage and shown the enemy his powerful weapons and fierce intent. It was to hop, skip and jump through clusters of reporters for the purpose of singing their boss's praises. Karen Hughes exulted: "I just talked to my husband on the phone, and he said, 'I'd like to see how the media can write that the governor didn't win this one.' "

Meanwhile, Bush's aides increased their efforts to find showcases for Bush that would make him seem bigger and better, and Bush's remarks in some of these places were not part of a diabolically clever attempt to be underestimated; they represented a concerted effort to be estimated more respectfully. His advisers felt—rightly—that he was more relaxed and voluble in small groups or individual interviews, so in late December, they booked him on CNN's *Larry King Live*, a station of the cross for presidential candidates. At the last minute, he consented to double the time he would give King, from a half hour to a full hour.

And he said something that, for the diligent Bush-watcher, was fascinating in the extreme, a clear indication that he was not really so content to be pegged as a puny thinker, that he was recognizing the trap of his rap. Asked by King to offer a rebuttal to all the pundits and political operatives who had portrayed him this way,

Bush said, "They ignored the fact that I went to Yale and Harvard," the latter being the university at which he earned an M.B.A. in the early 1970s.

Here was a politician who purported to abhor name-dropping, who ceaselessly endeavored to cast himself as a humble and people-smart Texan and curled his lip at anyone overly identified with elite enclaves in the Northeast, and he was directing voters to his ivy-draped diplomas. Necessity had become the mother of reinvention.

CHAPTER THREE

AROUND THE TURN OF THE YEAR, THERE WAS A lot of talk emanating from the Austin headquarters about the need to present Bush in a new, more assertive light, to pump up the volume. He was going to be less cautious, more combative, underlining and hammering away at the differences between him and Forbes and him and McCain on issues like education and tax cuts. He was going to mix it up more with voters, like McCain was doing—hold more town hall–style meetings at which his stump speech gave way more quickly to extended question-and-answer sessions with voters. He was going to show people just how deft and substantive he could be.

There was only one problem with the plan—Bush. He improved somewhat on the stump, although there were awkward moments like an exegesis on Middle Eastern policy that culminated in the profound proclamation that Israel remained "our buddy." He improved somewhat in debates, largely by unshackling his sense of humor and using it against his opponents. But an unusual theme started coursing through his less scripted remarks. He talked about how much he missed his three cats, Ernie, India and Cowboy, and his dog, Spot. He talked about how much he missed his morning ritual at the governor's mansion in Austin: waking up with Laura, padding downstairs to feed the pets and make the coffee, then bringing a couple of fresh cups, along with the morning newspapers, back to the bedroom so he and Laura could read in bed. He talked about New Year's Day on his beloved ranch near Waco, Texas, and how eager he was for some down time when he could get back there. Bush, it turned out, was homesick, so much so that he could not help but let it show.

It was a remarkable confession, because he had not, up until January, kept a schedule that required him to spend many weekends outside Texas; he had never gone more than six or seven days without circling back there. Apparently, it was an old habit: When he had worked full-time on his father's successful 1988 presidential campaign, he had special dispensation to interrupt an extended West Coast swing and, at considerable cost to the organization, fly back to Washington to visit Laura and his daughters for twenty-four hours or so before reconnecting with the troops in the field. Now, as he waged his own presidential campaign, his first extended spell on the road was getting to him. Could he not have realized how grueling it would be? Could it really have been such a surprise?

Amazingly, yes. Just hours after his virtually preordained victory in Iowa, where McCain did not compete and Forbes's lavish investment merely confirmed what the Beatles had said about money and love, Bush left for New Hampshire on an overnight flight that suggested an effort to keep a brisker pace. But along the way, he could not stop referring to how depleted he felt and how desperately he craved a break. Maybe the setting had loosened him up: instead of a big plane that could accommodate the entire Bush press corps, he was using a smaller jet with just sixteen seats, four of which were given to Glen Johnson of the Associated Press, Wayne Slater of the *Dallas Morning News*, Keen of *USA Today* and me. He knew all of us fairly well, and had picked us partly for that reason. As our tape recorders rolled and an entire hour went by, he sounded less like a man buoyed by excitement about his progress than a man beaten down by the rigors of what he was doing. He wanted a ticket home.

It was pitch-dark outside the windows, the engine hummed loudly, and he leaned in close to us, aware that his voice might not register so well on our recorders amid all the white noise. He looked weary, even though he had taken a long nap that afternoon as he waited for the Iowa returns, his anticipation of the results proving no contest for his desire to sleep. Such snoozing was a Bush leitmotif: back in June, when he concluded his first campaign swing at his parents' summer place in Kennebunkport, Maine, and held a news conference there with his father, the former president continued taking questions from reporters as his son drifted inside for a midday slumber. Later, during the day of the South Carolina primary, the Texas governor would doze even as the first exit polls came in and his father placed frequent, frantic calls to campaign officials for the latest updates. If Rip Van Winkle had sought higher office, his candidacy might have looked something like Bush's.

"Today I was exhausted," Bush said on the flight from Iowa, in a voice that supported this description. "I was dead tired." He conceded at another point that "this last week was hard. We were on an airplane every day, traveling, getting back late at night, and it was strenuous. It took a lot of energy." Bush had just racked up his first formal success of the Republican primary season, had just tasted triumph for the first time, and he was emitting not whoops of excitement but sighs of exhaustion. And not just a few of them, but one after another. He could not say enough about how wearying this whole ordeal was. Looking ahead to the next week in New Hampshire, he said it would be "a test of stamina" that would require him and other candidates to "dig down into our reservoir of energy." Looking beyond New Hampshire, he said, "I'll tell you where my sights are. What I'm really looking forward to is having a long weekend in my own home."

I'm not sure that any of this hurt Bush, in a political sense, one iota. In some ways, it reflected well on him. It confirmed for voters that his desire for the presidency was not so overwhelming that it obliterated other needs and concerns, and this would eventually play nicely against the perception of Gore as someone so intent on success that he would shift shapes and betray his principles to achieve it. Americans could be wary of too much ambition in a leader and too sharp a craving for popular acceptance, and Bush was not stirring such suspicions. But there was always a danger of going too far in the other direction, and Bush's ignorance or dismissal of that—at least until he felt utterly compelled to stop ignoring or dismissing it completely—suggested a certain honesty and self-acceptance. In time, those qualities would turn out to be hugely valuable to him, sources of steadiness during unsteady times and, especially after the terrorist attacks on the World Trade Center and the Pentagon, ways to connect with Americans in an appropriately emotional way. On some level, what Bush could give them was a kind of personal authenticity, and it made up for much of what he had a harder time delivering.

But in terms of his presidential campaign, the finite boundaries of his tolerance for the grueling, time-consuming tasks at hand underscored anew just how oddly matched Bush and his chosen goal were. They also posed a kind of logistical problem: psychologically and physically, could he go through all the paces that he had to? Could he make peace with the fact that there was no way to avoid them?

That night on the plane from Iowa to New Hampshire, I asked him whether, to make his journey more bearable, he kept anything special with him or made any adjustments to his hotel rooms. It was then that he told us about his "feather pillow." Brought from home, it went everywhere with him; it even had its own bag, toted by his personal assistant. This pillow had a durable half life in news media reports, for good reason. It was a remarkably telling metaphor for his yen for comfort and lust for the familiar, traits that came through in ways small and large, insignificant and potentially significant. They were diametrically opposed to the very nature of a presidential campaign, but they were essential to many of the decisions Bush made, and they partly explained the way he had lived to this point.

Despite his wealth, his father's experiences trotting the globe and periods of his life when he had plenty of free time, he had done relatively little traveling outside the country, at least until his mid-forties; he stuck to the places he already knew. He liked what felt cozy and safe: as the campaign progressed and he talked to those of us covering him about the new house that he and Laura were having built on his ranch, he said he was letting her decide on most everything, with a few exceptions. He wanted comfortable couches. He wanted a comfortable bed. He wanted a really good, comfortable shower. He wanted plenty of places for friends to stay, because he liked to be surrounded by the friends who had been constant through the changing phases of his life.

Much further along in the campaign, when he was deciding where he would spend the "down" days during the Democratic National Convention and the swerving of the media spotlight away from him, he rejected the idea of going on a traditional vacation to a faraway place. "We started thinking about a lot of places to go," he told those of us on his campaign plane one day, referring to himself and Laura. "I said, 'I want to go to my own place, my own bed, my own house, my own dog.' So that's what I'm going to do." He was bound for his ranch.

On the first day after he ended and emerged from this respite, he hastily rearranged his campaign schedule and left Tennessee for Texas several hours early, because he was due to take a run and decided that he wanted to do this not in Memphis but in Dallas, on a track he had used countless times before. He preferred the old to the new, the tried to the untested. He ate the same foods over and over, stocking his plane with peanut butter because that was his favorite sandwich ingredient. He did not like to have even this routine disrupted. When McCain, having lost the nomination to Bush, accompanied him on a campaign trip in the summer of 2000, one of the senator's aides boarded the Bush plane before Bush did, saw a peanut-butter-and-jelly sandwich lying around and, thinking it was communal food, ate it. Bush later went looking for it—and for the culprit who had effected its disappearance. "Who ate my peanut-butter-and-jelly sandwich?" he growled. His displeasure was ill concealed and less than tactfully conveyed.

The nicknames and cheek-pinching were not just about control and incorrigible (and platonic) flirtation; they were about taking what was unfamiliar and rendering it instantly familiar, turning a potentially threatening atmosphere into an unthreatening one, putting himself at ease. Similarly, his almost obsessive adherence to the daily campaign schedule that had been laid out for him and his famous punctuality were not just about politeness; they reflected a desire to make his world as predictable—and manageable—as possible. One of the reasons that the terrorist attacks provided such a particularly interesting challenge for Bush was that he had frequently seemed less interested in expanding his horizons than in drawing them to his specifications and less keen on opening himself to surprises than in making sure there weren't any.

He surrounded himself, whenever he could, with people he had known long and well, people whose tics and tendencies had often been brought into compliance with his own. Many political analysts were thunderstruck when Bush put his presidential campaign in the hands of three senior advisers—Rove, his chief strategist; Hughes, his communications director; and Joe Allbaugh, his campaign manager—from Texas and later anointed Don Evans, who had virtually no real political experience, his campaign chairman. One of Evans's principal qualifications was that he went all the way back to Bush's start in the oil business in Midland in the mid-1970s and was among Bush's closest friends. But Evans, like Rove and Allbaugh, was in sync with the rhythms and rituals and banter of Bushiness. With an arresting uniformity, the men around Bush had all come to mimic and master his minimalist patois, in which "Hey, buddy" or "How ya doin', buddy" replaced "Hello" and in which the all-purpose "You're/He's a good man" came to connote everything from genuine integrity to an ability to chew with your mouth closed. Bush didn't necessarily conform to the worlds he entered; he made them conform to him.

This explained his resistance, except when it became absolutely necessary, to conducting too many town-hall meetings: anything could happen in these forums, and at any moment someone could loft a question, criticism or remark that buffeted the discussion into uncharted waters. Later on, Bush's selection of Dick Cheney as a running mate was, in part, a triumph of safety over serendipity, as were Bush's subsequent selections, once he was elected president, of so many other members of the old Republican guard for cabinet positions. They did not necessarily make good on his promise to give the party and its politics a fresh spin and a new twist, but they were known quantities. And Bush was a creature of habit, determined—no matter what else he would have to endure—that the cushion beneath his head at night fit and feel exactly the way it always did.

When the feather pillow came up on that night flight from Iowa to New Hampshire, he said something else that was equally arresting. He asked us if we too traveled with our own pillows, and expressed surprise that we didn't. It reminded me of a conversation I had had just a few weeks earlier with Clay Johnson, who was then Bush's chief of staff in the governor's office and had been a close friend of his at Andover, where they were in the same class, and Yale, where they were also classmates. Johnson had been praising Bush's discipline, which was perhaps a positive spin on Bush's adherence to routines, but Johnson seemed truly impressed by it. He cited as an example Bush's typical workday as governor: Bush got to the office around eight or eight-thirty A.M., left around five-thirty P.M. and, at the midpoint of that stretch, took a break of ninety minutes to two hours—usually for exercise, sometimes just for a quiet respite. It was his "personal time," Johnson explained, and it was impregnable.

"He had two hard half days," Johnson said, rather than a full, run-on, unwieldy one. "It requires a lot of discipline not to get sucked into a lunch or a meeting or something. But it's an inviolate part of his routine, and so, be it a light lunch or exercise or whatever it is, he insists on getting that." Johnson added, "It's very hard to do, as you can imagine. All my friends say it's impossible to exercise. I say, how can the governor carve the time out, if you can't?"

How can the governor carve the time out, if you can't? Easy—he's the boss. There is no one above him; his schedule is the one to which everybody else's must bend; and he can delegate as much (or as little) as he deems appropriate. He can do almost anything he wants. It's like being pope, only without that nifty little bubble-car and the sky-high miter. Johnson's notion that Bush deserved special credit for—and faced special obstacles in—preserving the sanctity of his midday break flipped logic on its head, and if Johnson had come to see it that way, it was probably because Bush did, too. Like the assumption that travelers brought their own bedding with them, it made Bush seem disconnected from reality and coddled.

To some extent, he was. Before Bush ran an unsuccessful 1978 race to represent the Midland, Texas, area in Congress, he attended a "candidate's school" set up by the Republican party. David Dreier, who was there with him and later went on to become a congressman from California, remembered that, one day, Bush announced a brainstorm. "I've got the greatest idea of how to raise money for the campaign," Bush told Dreier, his voice brimming with excitement. "Have your mother send a letter to your family's Christmas-card list! I just did, and I got $350,000."

No sweat. No panic. Which, in many ways, described the way he spent the days before the New Hampshire primary. The idea had already taken hold that McCain was scrapping and scraping for votes while Bush was politely requesting them, that McCain was sprinting while Bush was gliding. Still, Bush reverted to events at which he kept detailed policy discussions to a minimum and glad-handing to a maximum, often spending twice as long signing autographs for voters as he did telling them why he should be president. He took generous midday breaks and staged a succession of photo opportunities devoid of any real content. He went to an indoor soccer arena, assumed the position of goalie and let children kick the ball at him. He went sledding down a snowy slope. He went candlepin bowling. He visited a snowmobile dealership and rode one of its wares in little loops out back. "This is better than a press conference," he told those of us watching him. "It's exhilarating."

He built some of his other events around high-profile endorsements, which essentially let other people do the work for him and sent the message: vote for me because all of these other people think you should. He stood side by side with former New Hampshire governor John Sununu, who had also been the chief of staff in his father's White House. He stood side by side with Jack Kemp, who had been the secretary of housing and urban development under his father and was still hurling that imaginary football. Then he brought in Poppy himself, staging an enormous rally in an indoor tennis center at which the old-time country-folk band the Bellamy Brothers played music for longer than all of the speakers combined gave the audience meaningful reasons to vote for Bush. Almost the entire Bush clan took the stage, and when former President Bush spoke, he said, "This boy, this son of ours, is not going to let you down," a statement that made the candidate sound like a fledgling and was widely derided in the press. The son himself talked for less than five minutes.

All of this fueled a sense that what Bush sought was a coronation, and all of it raised questions about how badly Bush really wanted the presidency or, at the least, how hard he was willing to work for it. When Bush himself had previously pondered his astonishingly rapid ascent in politics, blessed by good luck and great timing, he said he felt like a cork in a stream, buoyed by currents much larger than he was. It was a waterlogged metaphor that included no references to paddling or swimming or taking charge of his direction and velocity, and it was not at all clear that Bush was doing this now.

Reporters weren't the only ones who noticed. Many prominent Republicans found Bush's approach to New Hampshire to be half-hearted and presumptuous, qualities that might not prevent him from getting his party's nomination but that certainly had implications for the general election. They began to wonder—aptly but belatedly—if he had the gumption and grit to go the distance.

I told Hughes that I was going to write about this, and on February 4, as Bush traversed Michigan, which held its primary toward the end of the month, I asked her if Bush wanted to comment on the theme. She spirited me into a back hallway at a school in Detroit that he was visiting and put me face-to-face with him. "Make no mistake about it," he said. "Not only do I want to win, I'm going to win. I want to win bad enough to win, and I'm going to win. You can mark it down in your book: on this day, I told you, looked you in the eye, and said I'm going to win the nomination, and I believe I'm going to win the presidency." There was no doubt, no equivocation, no sense of alarm in his voice.

But his concern about my line of inquiry later became clear. That night, just before his campaign plane left Grand Rapids, Michigan, for Austin, he saw that I had my cell phone pressed to my ear and was talking hurriedly with an editor. I was going to be in the air—and out of contact with New York—for the next three hours, and my editor, Jim Roberts, was going over some last fixes to my article with me. Suddenly, Bush was all over me, smothering me with questions. What kind of lead had I written? What was the headline going to say? Was this some big front-page deal? Hearing all the disruptive chatter in the background, Roberts asked me who was causing the commotion. "Well," I told him, "that's the governor of Texas."

The article was published on the front page on a Saturday morning, and around midday, I had to call campaign advisers for another story I was working on. I tried one senior official, who was pronounced unavailable and never called back. I tried another: same deal. I finally reached one of the spokespeople, who told me, in no uncertain terms, that everyone was very displeased with me. In one sense, I was surprised: I had given them advance warning, and I had given Bush his say. In another, I wasn't. Bush was surrounded by a bevy of loyalists who didn't always see him clearly and were sometimes genuinely confounded that anyone could have a contrary view.

Campaigns often have aspects of a cult: the leader is all-powerful, all-knowing and everyone publicly defers to his or her wisdom. This was especially true of the Bush operation. His aides insisted that, in private, they were free to disagree with him and he was open to all advice, but they seldom gave any hint of this once they left the confines of the campaign headquarters. Almost to a person, they never referred to him as anything but "the governor," and this included even those advisers who were close social friends of his and even those occasions when they were dining or drinking with reporters off the record. "The governor believes," they would say, or "the governor has worked hard," "the governor decided," "one of the governor's most impressive accomplishments," "one of the most moving things the governor ever said to me."

He had constructed his campaign so that no one had any competing allegiances, any challenges to their fealty. Rove was instructed to sell his independent consulting business so that there was absolutely no distraction to his energies as the chief strategist. Rather than contract the campaign advertising to an existing firm, the Bush campaign created Maverick Media, led by Mark McKinnon, and its sole client was Bush. No policy that the campaign unveiled, even if it had been hashed out in a task force of prominent intellectuals, was presented as anything less than the governor's divine inspiration. It was a system and philosophy designed so that Bush would never be undercut or diminished. And sometimes it was downright spooky.

As the person with titular responsibility for the campaign's public face and actual responsibility for even more, Hughes was its duty officer and disciplinarian, keeping a careful watch over everyone in Austin and people far beyond. If you called one of the campaign's satellite advisers outside of Texas to talk to him about the presidential race, he typically asked if he could call you back in just a little bit—he was awfully busy right now—and then contacted Hughes to get clearance. This was obvious because when he finally returned the call, an hour or two or three later, his responses sounded rather curiously like hers. Or because Hughes herself would later show her hand and say, "I know you called so-and-so about . . ." If you asked her campaign subordinates an unanticipated question for which the person had not already heard and absorbed Hughes's preferred answer, he or she, too, had to get back to you shortly. Even after Bush won the election and moved into the White House, her vigilance didn't wane. An administration adviser told me that she had requested a seating chart for one of the big news media dinners that take place annually in Washington, so she could see which administration officials were sitting with which reporters, information presumably useful in determining the source of any future leaks.

She had a vivid wardrobe of green and lilac pants suits and a big presence, her six-foot-frame complemented by a booming voice. If Bush cracked a joke during a news conference, her laugh drowned out anybody else's; sometimes it was the only laugh around. She knew his mind and manner of speaking so well that when he answered questions, her mouth would sometimes move silently in concert with his, like a ventriloquist without enough stealth.

She was his most ardent fan, with a "Go Bush" license plate on her green Mazda 626. She was ceaseless in her promotion of him, known to trundle to the back of the campaign plane to share the *funniest* thing the governor had just said to her or the *sweetest* observation he had just made. And she was his fiercest defender, never willing to acknowledge any inconsistencies or flaws in his record, performances and positions. Terry Neal of the *Washington Post* and I once spent twenty minutes asking her to explain how Bush could say that he opposed abortion except in cases of rape, incest or danger to a pregnant woman's life and could also say he supported the Republican platform, which made no such exceptions. She looked at us as if we were bonkers and stated, over and over again, that the platform was an expression of support for the rights of the unborn and Bush saluted this. Yes, but what about the details, what about the discrepancy? Bush, she said, wanted to move the country toward a greater appreciation of life, and this was what the platform advocated, too. We asked the question a dozen ways and got nowhere. Hughes was immovable.

She could find a Bush-flattering version for just about any sequence of events. Shortly after he lost New Hampshire, headed to South Carolina and finally committed himself, day after day, to spending as long as sixty minutes answering substantive questions from voters, a reporter seeing Bush for the first time told Hughes: "I never knew, from what I'd read, that he did this." That was because he hadn't, not really, but Hughes provided a different explanation. She said that Bush had done this all through New Hampshire and even before; it was just that he did it after he left the microphone and voters approached him one by one, and the campaign subsequently realized that reporters were ignoring it because they couldn't hear it. Not quite. Bush was putting his arms around these people, fielding their compliments about his fine parents, smiling for their personal cameras and signing his name to their baseballs or other trinkets. More often than not, he wasn't talking about the standardized test scores in their schools or their troubles with their health insurance.

It stood to reason that Hughes would be upset about the article I wrote in February, with its suggestion that Bush needed to find more passion, more purpose and more stamina than he had thus far demonstrated. And, indeed, she was. She made her gripe publicly known that Monday morning, when reporters boarded the campaign plane in Austin and she saw that I wasn't there. Another reporter for the *Times* was spelling me for a few days so that I could get back to Washington, file overdue expense forms and do some reporting that was impossible to accomplish on the road. Hughes spotted my replacement and took off: was the campaign's supposedly sluggish pace too much for me? Was the fire in my own belly a mere ember? Had I not, during the final week in New Hampshire, spent a day and a half in my hotel room while a *Times* colleague took my place at events? (I had, but then I wasn't running for president, and when Bush was taking breaks in the afternoon or hitting the sack at night, I was often transcribing and writing about what he had already finished doing.) Before the plane left the ground, my cell phone chirped with one call after another from colleagues letting me know about Hughes's comments.

I sent her an e-mail, letting her know that I would have appreciated it if she had taken her complaints to me before publicly announcing them to my peers. That evening, she sent me a response, telling me that she had really been joking and noting what she felt was an irresistible irony, and I think we were both content to let the matter end there. But word of what had happened between us somehow reached Bush, and a *Newsweek* reporter subsequently overheard him ask Hughes: "Did you apologize to Panchito?" (This was his latest nickname for me, a derivation of "Pancho," one of the acceptable Spanish translations of "Frank.") Hughes responded, "Do I have to grovel?"

I rejoined the campaign several days later in South Carolina, and Bush's reaction was civilized and at least superficially amiable. At the conclusion of the first event at which he spotted me, he walked up to me, wrapped his arm around my shoulder and said, "You know we love you!" He saw no percentage in any enduring rift. He understood that the past could not be changed but the future was always up for grabs. He also liked the environment around him to be as happy and festive as possible, and reporters like me were part of the climate. At the next event, a question-and-answer session at a high school, one of the students asked him how he regarded the coverage he received from journalists assigned to his campaign. He usually answered questions like this by saying that he would love to change some of the articles but respected the independence and commitment of the press and even liked some of its representatives. This time, he went further, looked across the crowded room in my direction and ad-libbed: "I love you, man!" People in the audience had no idea what he was doing.

A part of me sympathized with what he was going through. The campaign trail presented an exhausting, discombobulating, tedious way to live. It was an exercise in redundancy and a recipe for madness, with a herky-jerky rhythm of plane trips followed by bus trips followed by cookie-cutter events followed by bus trips back to the plane.

At night, upon arriving at a hotel, those of us in the press corps had to wait up to an hour for our suitcases to catch up with us before we could go to bed. In the morning, usually ninety minutes or more before our scheduled departure, we had to deposit our bags in a designated area so that security agents could rifle through the contents, and bomb dogs could sniff and leave dirty paw prints all over them. This meant waking up ridiculously early, a situation that some reporters tried to circumvent by taking the next day's clothing and all the requisite toiletries out of their luggage before going to bed and never even moving the bags to their rooms. This cunning but imperfect strategy inevitably led to mishaps.

I twice tried this method and twice forgot a semi-essential article of clothing, and thus strode around one frosty morning in Michigan in black loafers with no socks. Bush noticed—he was keenly attuned to people's physical appearances, to whether they had skipped shaving or gained or lost a few pounds—and asked me if I needed to borrow a pair. I cordially declined. There were certain boundaries between candidate and reporter that didn't need to be crossed.

The Associated Press's Johnson could be seen one morning walking around a hotel lobby in Nashville, Tennessee, in a white T-shirt, gym shorts and the cowboy boots he always wore in the spirit of the candidate's home state. He had handed in his suitcase for inspection the previous night and had forgotten to extract his slacks, and now he was stuck: the luggage was already on its way, ahead of the buses, to the airport. He threw himself at the mercy of campaign aides, who gave him a special, private, advance ride to the plane so that he could reunite with his pants and be spared the humiliation of too many colleagues taking a gander at his gams.

We had to worry as much about our laundry as our deadlines, because we usually changed hotels nightly, which did not give us enough time to have our clothes cleaned, and none of us wanted to weigh down our bags with nine or ten days' worth of outfits. So we studied each week's travel schedule carefully, in the hopes of identifying a luxury hotel that promised a valet service with a four-hour, overnight turnaround. If we spotted such an opportunity, we quickly grew jubilant. The news traveled through the plane faster than a juicy, substantive political tip.

And we ate, and ate, and ate. In the mornings, we were provided with enormous breakfast buffets: eggs, sausage, bacon, potatoes, pancakes, fruit, cereal, yogurt. Then came another breakfast on the plane, a lunch buffet in the subsequent destination, more food on the plane, a dinner buffet at night and snacks at every imaginable interval in between. Campaigns live by the unwritten rule that a well-fed press corps is a docile press corps, and rare was the political operation that wanted to lose any points with reporters by failing to coddle them with calories. (Besides, the bills for all of this care and feeding were going back to the news organizations employing the reporters who were being cared for and fed.)

Arkansas meant enormous platters of barbecued ribs and barbecued chicken. New Mexico meant a miasma of guacamole and sour cream. In Oregon, we served ourselves from a bowl of Caesar salad that was bigger than a bathtub. In South Carolina, we were given individual Styrofoam containers heaped high with every kind of fish that had ever been deep-fried and some that had never known the ignominy of batter and oil before. I grew like the Blob, amorphous and infinite, and during breaks in campaigning, I found myself marching to the nearest mall for new pants with another inch or two in the waist. The seats on the plane seemed to get smaller and smaller as the campaign dragged on.

We were prisoners in a gilded cell, which we called "the bubble," because it was so separate from everyday life and had an atmosphere all its own. We were bound to printed schedules that dictated exactly how long we would stand in one place or sit in another. We had discreet windows of time in which to complete our stories before we were yanked from any and all access to the phone lines that enabled us to file to our news organizations, and then we were gone again, barreling down the highway or tunneling through the clouds. We were following, not leading—herded like sheep, lured like lemmings. Small wonder that we got so cranky. Smaller wonder still that we got so silly.

Alexandra Pelosi, a producer for NBC, was known to dance little jigs or burst into song or get on the public-address system in the front of a bus and pretend to be a cruise director. She carried her own personal video camera with her, turned it on and asked her colleagues offbeat questions, goading them into self-parodying soliloquies. Pelosi, the daughter of a Democratic congresswoman from California, was the unrivaled queen of the pack when it came to self-amusement and consequences-be-damned diversion, taking advantage of empty spaces in a day's schedule to head to the nearest Kmart for festive new underwear, which she would then show us, or visit friends or relatives. In Houston, she had to catch her own taxi back to the airport from which the campaign plane was taking off, because she had lingered too long at the mall and the bus departed from an event at a downtown hotel without her. In New York City, she had to hire and ride in a helicopter from midtown Manhattan to LaGuardia because she was off running errands and seeing friends.

I developed a fixation on those little moist towelettes stacked in dispensers in some bus and airplane bathrooms. I would collect them, hoard them and then pass them out among colleagues, chattering about the virtues of a wet, citrusy burst of facial refreshment as a remedy for that oily post-stump-speech sheen. It was self-conscious shtick, designed to pass the time, and it even ended up on an MTV segment about the boys—and girls—on the bus. I kept wondering when the makers of Wet Naps would call and offer me an endorsement deal. It was my fantasy of a way out.

Other reporters had other indulgences. For several weeks, John Berman of ABC, T. Christian Miller of the *Los Angeles Times* and Kevin Flower of CNN went bonkers for a game on their PalmPilots called "Dope Wars," the object of which was to make the most money in an imaginary version of the drug trade. They could be heard to shout out dollar figures or whoop in triumph whenever they scored an especially lucrative sale of heroin or cocaine.

A glamorous life? We stayed in one motel beside a stretch of railroad tracks that seemed to bring a new, screaming train every hour, all through the night. We stayed in another where the carpets were palpably wet with mildew and the smell corresponded to that reality. We battled constant technological crises: the cell phone whose battery died at the most inopportune time; the laptop that froze every twenty minutes, erasing the last fifteen minutes of work done on it. Richard Wolffe, a reporter for the *Financial Times*, set his computer bag down on the pavement one day and then looked up in horror to see the rear wheels of a campaign bus backing straight over it. Crunch, crunch; bye, bye. That was all, at least on that day, that Wolffe wrote.

Hard-pressed to keep track of all the cables, adapters, gadgets and whatnots in my own bag, I twice lost my computer's power cord, once before a particularly significant day, and spent the hours between seven and ten P.M. scouring malls around Pittsburgh and placing frantic calls to the IBM help line. I came up empty. My friend Patsy Wilson of Reuters saved me, because she had the same equipment and was willing to sit next to me and pass her power cord back and forth, as if we were scuba "breathing buddies" using the same oxygen supply. By the end of the campaign, I had gone through five different laptops, each of which met its own quirky demise. One was dropped, one was stepped on (yes, I realize I am using passive constructions here) and others simply quit, probably in protest of all the wear, tear, use and abuse. They were smarter than I was. They had no interest in going the distance.

The most ridiculous things took on the greatest importance, given the dreariness of our lots. Before the campaign assigned the seats on the plane, small groups of us banded together for the sole purpose of securing and guarding aerial real estate in the exit rows, which meant extra leg room and also happened to correspond to the spot in the aisle where Bush was likeliest to stand whenever he drifted back from first class. These charter flights were their own freaky universes, governed by strange rituals and bizarre doings. It is a long-standing campaign tradition for reporters, as the plane takes off, to find an orange, scrawl a message on the rind and try to roll it, like a bowling ball, uphill to the candidate, who writes his own message and sends it back. But this particular fruit was not always available, and there were times when Bush, beginning the exercise on his end, attempted to send a pineapple or melon in our direction, which led only to it swerving away from the aisle and under a seat during its journey.

A kitchen fire broke out at one point, sending smoke through the plane; the flight attendants dismissed any concern, saying that this happened all the time. We once idled on the tarmac for more than a half hour because one of the plane's eight wheels was defunct and had attracted the attention of mechanics; we ended up taking off anyway, assured—but not wholly convinced—that the remaining wheels would do. And our landings, well, I don't know what it is about charter pilots, but they seem less attentive to the niceties of impact than commercial pilots do. We swerved, teetered and bounced, our hearts in our throats, our stomachs in tatters. I swear I had a near-death experience at one point—tunnel, light and all—but it was probably just a bad carrot from the crudite. I was doing roughage that day because I couldn't face yet another trip to the mall.

None of this was incidental to the coverage that we provided voters; our universe was so insular, so peculiar and so mundane that we inflated the importance of things, insisted on finding friction where it was not readily apparent. Some reporters' stock in trade was to goad one candidate into criticizing a second candidate's position on an issue, use that comment to coax a retort from the second candidate, then write about how the two had "feuded" or "sparred" on the campaign trail that day. Of course the candidates had—they were set up to do so. And if you ignored it, you faced—or just as likely imagined—the threat of questions from editors about why everyone else was noting something you were ignoring. So you followed suit, or at least I did, more often than I'm proud to admit, because it was easier and livelier to write about an argument than it was to write about anything else. The resulting stories sometimes skimped on the actual substance of the candidates' differences on the issue, because this information, the most potentially useful to voters, had often been reported at some point in the past. It wasn't "new," an adjective sadly estranged from "meaningful." So you put on your pinstripes and joined a game of insider baseball that always went into extra innings.

Bush knew the score. And what he started doing after New Hampshire was what he had failed to do there: he got up to bat more often and swung harder at the pitches. His public graciousness about his New Hampshire defeat and his advisers' portraits of his equanimity were only half of the story; at a dinner of family and friends when the returns were coming in, he was sullen and angry. He had been mortified and knew it, and both a pride that had been dormant and a humility that had been absent kicked in. The length of his days on the trail grew, as did his animation at events. He strayed more frequently from his script. He wasn't suddenly, or even remotely, terrific on the stump, because he still had his speech impediment and still landed on the goofy side of colloquial. But he was certainly trying harder, and that addressed a core worry about Bush—a possible flaw—that was more pertinent than his book learning or manner of speech and would become increasingly so once he entered the presidency and confronted the crisis he did. Did he and could he care enough to commit and apply himself fully when the chips were down? Could he be as purposeful as the most urgent circumstances might call on him to be?

South Carolina was like a first little laboratory for that inquiry, and while the results weren't pretty—Bush played rougher and more partisan than he ever had before—they were somewhat promising, because Bush was at least snapping to attention and playing for keeps. Right after his flight from New Hampshire arrived in South Carolina on February 2, he appeared and gave a speech at Bob Jones University, a conservative Christian school with racist and anti-Catholic leanings. This was a temple of the religious right, which would be crucial to success in South Carolina, and Bush knelt before it. He would later say, in defense of himself, that he had gone there because it was important to bring his message of inclusion, tolerance and compassion to everybody, and that he had spoken out against the school's ban (which was later lifted) on interracial dating. This was, at best, a quarter-truth: his repudiation of the ban came at a subsequent news conference at a different location, and was issued only in response to a challenge from one of the reporters present.

Around the state, his supporters talked considerable trash about McCain, tarring the senator as mealy-mouthed about abortion and morally suspect. A Bob Jones professor sent out e-mails falsely claiming that McCain had sired two children out of wedlock. Invisible hands distributed fliers attacking McCain's wife, Cindy, for her addiction to painkillers a decade earlier and her admission that it had prompted her to steal medication from a clinic where she worked. Unidentified voices placed phone calls to let voters know that McCain was on his second marriage. Bush and his aides said they had nothing to do with it, and there was no proof that they did. But whenever Bush was questioned about these hardball tactics, he did not respond with a full-throttle exhortation that they cease and desist. He instead offered a woe-is-me testimonial that he too was being attacked, by groups like the Sierra Club and the National Abortion and Reproductive Rights Action League. And he didn't bother to point out that these organizations were targeting his record, not his personal life.

He also strayed across the line himself, organizing an event on February 3 at which an eccentric veteran with a history of controversial statements charged that once McCain, a former navy fighter pilot, had returned from Vietnam, he abandoned his concern for the military and veterans. It was pure tackiness, given that

McCain had spent five and a half years as a prisoner of war there, and Bush stood mum beside the veteran throughout it. At other events, in a more justified vein, Bush questioned McCain's credentials as a campaign-finance reformer, noting that the sources of the senator's money and some of his conduct as chairman of the Senate Commerce Committee contradicted his holier-than-thou claims. This produced one of Bush's more illustrious one-liners, an insistence during a news conference just before the South Carolina primary that McCain "can't have it both ways. He can't take the high horse and then claim the low road."

It was in South Carolina that Bush visibly dusted off his competitive instinct, honed in a family that turned everything into sport. (At reunions in Kennebunkport, they staged ersatz "triathlons" consisting of tennis, horseshoes and golf.) He gave his most resourceful and animated debate performance in Columbia, South Carolina, flew to Hilton Head around midnight and was up at the crack of dawn for a morning event at which he handily dispensed with a sometimes bizarre array of questions and talked much more than he mugged. Perhaps none of this ultimately mattered as much to the outcome of the primary, which Bush handily won on February 19, as the way the state leaned toward establishment conservatives, a phrase that described Bush much better than it did McCain. But it showed that Bush could be made to focus in a crunch.

Not that South Carolina put any dent in Bush's astonishing capacity for detachment. Almost a year later, several people familiar with the story told me about a fixation that Bush developed on the night before the primary. When Bush walked down the hallway to his room in the Greenville Hyatt in the waning hours of February 18, he spotted an attractive brunette slip into the room of one of his campaign officials, Matthew Dowd. She was wearing a baseball cap, pulled low, and turned away from Bush as he spoke to Dowd, sparking Bush's curiosity. Bush then placed a phone call to the room of Karl Rove to ask if Dowd's wife was traveling with Dowd. Rove didn't know. Bush asked him—in a manner more voyeuristic and gossipy than judgmental—to find out, then placed a series of subsequent calls to Rove to needle him for the information. (Dowd's wife, it turned out, was indeed the mystery woman.) These were do-or-die hours for Bush, who had just been trounced by Senator John McCain, an Arizona Republican, in New Hampshire and could lose his grip on the Republican nomination if McCain repeated the feat here. But Bush was focused on whether one of his deputies was having a little naughty extracurricular fun.

CHAPTER FOUR

IT WAS EARLY MARCH, AND BUSH WAS PRACTICALLY lactating. Breasts were the topic of the day. Bush supported them. He was a little hazy on the proof of this—asked about the extent of Texas's efforts to fight breast cancer, he admitted that he did not know if there was money in his state budget for cancer research centers and said he would "have to check" on that—but there could be no doubt of his commitment. After all, he had a little pink ribbon on the lapel of his suit. He was surrounded by medical experts at a table inside a university that was known for its research into treatments and cures for breast cancer. He had even dusted off Elizabeth Dole, his onetime rival for the Republican presidential nomination, and plunked her beside him. She had breasts.

Bush was reconvening with his feminine side, his compassionate womb, because he was in New York, which held a primary on

March 7, and these kooky Northeastern Republicans favored a brand of conservatism that mixed testosterone with at least an equal portion of estrogen. Bush obliged, at least in the realm of semiotics. He did not carry any significant new proposals to this "panel discussion" in Stony Brook, on Long Island, where a high incidence of breast cancer had brought this issue to the fore. But he adopted a softer tone and geographically relevant visual devices for new political territory. Mars embraced Venus, at least until he encountered a state where Mercury or Pluto would get him better play.

Much was being made of Bush's political gyrations during this stretch of the primaries, of the notion that he had bent to the right in South Carolina and then angled toward the center in Michigan, which he nonetheless lost to McCain on February 22. Bush shot down this idea by pointing out that his positions on issues had not changed, and he was mostly right on that score. But there was another, equally significant tilt taking place: over the course of the campaign, depending on what he needed to accomplish, Bush's props constantly shifted, as did his stage design, his costumes, his supporting cast, the diction and cadences of his dialogue and the kinds of neighborhoods in which his act was playing.

Bush and his aides lifted political theater to new, Sondheimian heights, and demonstrated in the process how much could be fixed with powder and puffery, how thoroughly a candidate could be transformed from the outside in, how little he had to do but stand on the right set, under the right lighting, and say the right lines. If it was hard to figure out exactly what Bush was made of—and if, by September 11, 2001, it was not a whole lot easier—this was a good part of the reason. Politics had become as much about the show as the showman, and it seemed entirely possible that a candidate could prevail in the end not because he had survived a rigorous vetting, his mettle shining through, but because he and his stage hands had put on a fabulous production. The Bush campaign certainly did that, and what made their efforts all the more effective was that they had a candidate whose very lack of bold definition—whose spongy failure to make an emphatic mark—allowed him to assume the attributes of the scenery around him. It enabled him to be whatever people were inclined or wanted to see, a Rorschach running for president.

He struck different poses at different times. Early in his campaign, when he needed to establish an appropriate distance from Tom DeLay, Dick Armey and other grim-faced, dour-voiced, stiff-haired curmudgeons running the House, he did it linguistically and stylistically, with a few well-placed, well-timed phrases and a plethora of visits to black and Hispanic communities. He called their assault on the earned income tax credit an attempt to "balance the budget on the backs of the poor." Days later, he gave an education speech in which he railed against an outmoded Republicanism that equated the health of the country with the ripeness of its economic indicators—this was before he used the same measures to argue for his tax cut—and dismissed the Republican lament about an America slouching toward a latter-day Gomorrah. For added effect, those remarks immediately followed a visit to a mostly black charter school in Harlem, and several black advocates of charter schools flanked him on the stage as he delivered his address.

Bush and his advisers knew that to prevail in the general election, where the spoils would likely go to the candidate who most firmly occupied the political center, they had to soften any of the rough edges that many voters had come to associate with the Republican party. This effort hinged largely on arranging events, fashioning photographs and making appearances that put minorities front and center. It went all the way back to his announcement, in March of 1999, of an exploratory committee for his presidential campaign. Except for Laura Bush, the two people standing closest to him in pictures from that news conference were both black: J. C. Watts, a Republican congressman from Oklahoma, and Condoleezza Rice, the former provost of Stanford University and future national security adviser.

If Bush went to a school with a diverse student body, the children situated closest to him—the ones who would have to be included on television or in newspapers and magazines if the journalists recording the event widened the frame—were almost invariably black and Hispanic. He prayed in black churches. He went to Central High School in Little Rock, site of one of the most famous battles in the South to integrate schools. He embraced no past Republican president as tightly as Abraham Lincoln, who put an end to slavery. The apotheosis of this trend eventually came at the Republican National Convention in late July, where the stage featured nearly as many black performers and speakers as the Apollo Theater in a good month.

Almost everywhere he went, Bush's comments struck a canny, even-handed racial balance or made ideological allusions that went in several directions at once. He told children in Seattle that two of his heroes in life were Winston Churchill, a patrician white guy, and Willie Mays, a trailblazing black athlete. In a speech about health care, he managed to refer to former president Theodore Roosevelt, whom McCain had adopted as the patron saint of political reform; Pope John Paul II, the titular leader of all those coveted Catholic voters; and Booker T. Washington, a black educational pioneer. The man was covering all the bases.

Even the tape of music that was often played at the end of his campaign events, as he mingled with the crowd, seemed like a triumph of painstaking political calibration and aggressive image-making. In a nod to Hispanics, Gloria Estefan sang "Get on Your Feet" and Richie Valens crooned "La Bamba." There was country-and-western for the rural vote, a dash of Sting ("Brand New Day") and Van Morrison ("Bright Side of the Road") for the more sophisticated urbanite or suburbanite, some rhythm, some blues and some soul. John Fogerty's "Centerfield" evoked Bush's days as a managing general partner in professional baseball, the all-American sport. For a while, Tom Petty's "I Won't Back Down," an ode to tenacity, was in the mix, then it abruptly disappeared; an aide whispered something to me about the campaign never having acquired the requisite permission, which Petty's representatives were apparently unwilling to give. Another aide had suggested that Sheryl Crow's "A Change (Would Do You Good)" be put into rotation, but a careful review of the lyrics had revealed an unfortunate reference to self-gratification, and there was, presumably, no identifiable constituency that put this practice at the top of its political priorities.

Bush was not the only candidate giving good show. I will not soon forget the last day before the New Hampshire primary, during which the various contenders put any and all high-minded appeals in the closet and stepped out in whatever garb gave them the best shot at television time. (We in the media are guilty co-conspirators in this all.) Bush and Gary Bauer, the elfin and ill-fated protector of all things fetal, attended a Bisquick-sponsored presidential pancake flip-off, which submitted them to a leadership-defining test of their dexterity with a spatula. The goal was to send a pancake into the air, watch it do more than a dozen revolutions and then catch it as it headed home toward the griddle. The judges gave Bush a score of 59 out of a possible 60.

When Bauer's turn came, he lofted the pancake, took a step back and plummeted from the rear of the stage, vanishing from sight. It was a shame, because he had begun the event so promisingly, musing that the reason Forbes had not shown up was because he was "at a Fabergé egg scrambling contest" across town.

As for McCain's hyped and heralded bus, the "Straight Talk Express," which allowed reporters a kind of constant access to him that they did not get with other candidates, it owed at least as much to cunning political artistry as it did to a brave, risky stand for unimpeded journalism. He clearly realized what Bush would later come to see: that the danger of showing more of himself to journalists might, in certain circumstances, be outweighed by the benefits. That every time reporters wrote a story—and, in some cases, they wrote their stories just several feet from McCain, as the bus rolled along and McCain continued to chatter—they knew they would have to face him, for hours on end, after it was published or aired. That their proximity to him would make them feel important. That his generosity in giving them so much material and so much time would win him a degree of favor that was bound to inform their coverage.

One journalist who toggled between the Bush and McCain campaigns grew so enamored of McCain, and ate such a steady diet of invective about Bush from McCain's aides, that he called me one night just before the South Carolina primary to rant about how evil-spirited Bush was. He was in a bar with McCain reporters and McCain aides and it sounded like he had been there for quite some time. "Bush is the guy in the fraternity house who, when one of his brothers committed a date rape, helped cover it up," he told me. I thought for a moment that he was giving me a preview of a story he was working on, and asked if he was about to run with this. "No," he said. "I'm just talking about what he's like." It was one hell of a metaphor, and not the kind of statement you ever heard a journalist say about the beloved McCain.

In fact, you seldom heard them say much that was bad. Careful reflection on McCain's hyper-activist foreign policy and "rogue state rollback" program, under which the United States would dramatically ramp up efforts to arm and finance rebel forces in any country with a leader we found noxious, could be more than a little scary, but few reporters sounded any alarms. A careful listening to McCain's discussions of domestic issues other than campaign finance reform revealed a politician who, despite nearly two decades in Congress at that point, was still figuring himself out, but only a handful of journalists carefully examined this. And most of them, along with their editors, gave McCain a pass when he referred to the Vietnamese he had once fought as "gooks," a slur that might have left any other candidate reeling from news media criticism for days or weeks.

One of the reasons was that McCain was putting on his own brilliant show. He had released an autobiography that detailed his war heroism just as his campaign began, which meant that there were plenty of available, just-read copies for understandably impressed fans to tote with them to his campaign events, where he autographed the volumes. Some of these fans were even given tours of the "Straight Talk Express," which McCain's advisers had smartly turned into a living museum on wheels.

Contrary to myth, McCain was not available to *all* members of the news media; the size and configuration of the bus made this impossible. It was divided into two sections, the front of which accommodated his staff and wife and the back of which had room for McCain and no more than nine or ten representatives of, for the most part, the largest and most influential news organizations.

Representatives of smaller organizations sometimes got to rotate in, but not, for a significant stretch of the primaries, if the organization in question was the *Arizona Republic*, the most important newspaper in his home state. McCain hadn't liked its past coverage of him and spent part of his campaign pretending that it didn't exist. The reporters who were allowed on the bus squeezed into corners and sat cross-legged on the floor while McCain, at the center, stretched out in a recliner. He looked like a king on a throne with subjects sprawled at his feet, but no one ever described the scene this way.

The Bush campaign, in any case, was by far the most intricately orchestrated of the bunch. The whole operation was a symphony of subtle arpeggios and less subtle crescendos, and what Bush had done in South Carolina was insert some new bars and notes. He did not overtly advocate anything that he had previously repudiated or ignored, but he tinkered mightily with the delivery and scenery. At Bob Jones University, there were no black or Hispanic people beside him, and the person who introduced him, former South Carolina governor David Beasley, recited a litany of legislation in Texas that proved Bush's anti-abortion credentials, emphasizing an aspect of Bush's record that the campaign had seldom put in bright lights before. Beasley also noted, helpfully, that Bush "deeply loves the Lord."

Right after this appearance, he arranged to be joined at a news conference in Greenville, South Carolina, by Dan Quayle, a politician still popular among the religious right, still revered for his assault on what he felt was the television character Murphy Brown's glorification of single motherhood. Quayle pointedly and repeatedly used the word "values" in his remarks on Bush's behalf. As he did so, Bush, standing a few feet to his side, did something he normally didn't: he pulled Laura close to him and, at one point, held her hand. It wasn't that the Bushes seemed to lack affection for one another; they just seldom demonstrated it in public in a physical fashion like this. Coincidence? Perhaps. Or perhaps Bush was lending a well-timed, concrete illustration to Quayle's implicit claim that Bush was a solid, sturdy, traditional family man who would never be caught in the clutches of an intern. South Carolinians grooved on that sort of thing.

Up until South Carolina, Bush had brushed by questions about gays as quickly as he could, and what he did say mingled tacit tolerance with a tempered plea that society not get riveted by issues that put people in separate camps and fostered division. But the southern clime precipitated sterner words. In an interview on a Christian radio station, he suggested that he would be unlikely to appoint anyone openly gay to his administration. "An openly known homosexual," he said, using a phrase with sinister, disapproving overtones, "is somebody who probably wouldn't share my philosophy." He knew better: there were openly gay people in Washington and elsewhere, including a relatively close family friend of his, lending him unwavering and vocal support. In a subsequent debate, he addressed the subject of why he had not agreed to meet with the Log Cabin Republicans, a gay organization, by saying they had already officially signed on with McCain. He may well have known better in this case, too. The group had done nothing of the kind, but by insinuating that it had, he planted the idea in South Carolina voters' minds that McCain was too tolerant to be trusted.

Even some of Bush's Republican allies were bothered by this. Bush had started his race for the presidency by pointing to his overwhelming reelection margin in Texas as evidence of his ability, matched by his desire, to bring independents and Democrats into the Republican fold. But many independents and Democrats were flocking toward McCain, and the rules of the Republican primaries in South Carolina, Michigan and a few other places allowed them to participate in the voting. So Bush was now rallying and inciting the conservative base to turn out in droves and prevent the contest from being hijacked by outsiders. He used phrases like that.

After the McCain campaign placed phone calls in Michigan that reminded voters about Bush's visit to Bob Jones and gave them a little crash course in the institution's anti-Catholic transgressions, the Bush operation went through yet another semiotic overhaul. Suddenly, Catholic priests and Catholic themes were omnipresent. At a late-night rally in Kansas City, a collared clergyman stood near Bush on the stage. At a midday event in Springfield, Illinois, a collared clergyman sat near Bush on the dais. Bush managed, on a trip to upstate New York, to find a city that held a St. Patrick's Day parade two weeks early, and he zestfully joined it. He also managed, during a visit to a strawberry festival in Florida, to stop for a snack at a booth set up by a Catholic church, and the word Catholic was prominent on a banner in the news photographs that were taken of him that day. It would have surprised none of us keeping a close eye on Bush if he emerged on the campaign trail one morning with a heavy crucifix on his back, started cradling babies in the manner of the Pietà or replaced the crackers on his plane with communion wafers. He was in the midst of some serious atonement.

By the constellation of important primaries on March 7, he had exterminated McCain. The next constellation of important primaries on March 14 made it official: nothing outside of death could prevent Bush from becoming the party's nominee. It was grab-the-center time once again, and a victory celebration that he staged in Austin on March 14 was an ethnically, racially and politically polymorphous jubilee. The El Paso mayor who introduced Bush was both a Democrat and Hispanic. The campaign video that played on a large screen featured a sudden, incongruous and gratuitous close-up of a cleric, presumably Roman Catholic, who was smiling and laughing, presumably in adoration of Bush. The setting for all of this was a Jewish Community Center. But the Bush campaign's attention to such details faltered when it came to the buffet they initially planned to lay out for journalists. The selection of barbecued meats that the campaign had ordered included not just chicken and beef but something not allowed on the premises: pork. Learning this at the last minute, the campaign quickly deleted the item from the menu.

If Catholic priests were falling out of the trees, Democrats were not far behind them. There wasn't a prominent Democrat in Texas—where many Democrats, mind you, are considerably more conservative than Republicans elsewhere in the country—who wasn't pressed into appearing for Bush somewhere useful. There wasn't a single opportunity that Bush missed to talk about his chummy friendship with the deceased former lieutenant governor of Texas, Bob Bullock, a Democrat, whom Bush credited with nudging him to run for president. Bullock's widow, Jan, even ended up with a speaking slot at the Republican National Convention. Texas politicos confirmed that Bush indeed had Bullock's respect and affection, and would surely have had his backing in the presidential race, but said that Bullock's thoughts and energies didn't dwell quite so lavishly on Bush as Bush sometimes made it sound. Bullock, after all, had had an investment in good relations with Bush, who could sign or veto bills coming out of the Texas Legislature.

The Bush campaign also directed attention to its resident, in-house Democrat, Mark McKinnon, the head of Maverick, a Texas consultant who had never worked for a Republican until he met Bush, but was ostensibly so dazzled by Bush's sensitivity and forward thinking that he signed on with his 1998 gubernatorial reelection campaign. By 1998, of course, it was already clear that Bush might be the Republican Party's nominee in 2000, and McKinnon, like any political operative, wanted a ticket to the big game—wanted to play in it at least once. But this was not how he and the campaign marketed his story. McKinnon was served up to profile writers for newspapers and magazines as proof positive of Bush's ability to appeal even to progressive, arty, dyed-in-the-wool Democrats like McKinnon himself, who had once penned songs for Kris Kristofferson and favored a wardrobe that included a pair of chessboard loafers and a beret. He invited reporters over to the Maverick office, which was dimly lit and below street level and often had hip music blaring from a boom box.

Before South Carolina, any suggestions of a lack of fervor in Bush's opposition to abortion had been verboten. After South Carolina, Governor Christie Whitman of New Jersey, one of the most ardently pro-choice Republicans in the land, appeared on Bush's behalf in Northeastern states with a message meant to muddy the waters. She said that she supported Bush because he was a politician who didn't abide "litmus tests," a phrase used most often in politics to talk about the insistence by some politicians that administrative appointments or Supreme Court nominees be expressly pro-choice or pro-life. The faces at the places Bush visited grew conspicuously darker. The more pronounced Southern accent that he had brandished in South Carolina was gone; in its place were spasms of Spanish that signaled his solidarity with the nation's fastest-growing minority group. And a new, next-generation Bush began to show up at his side more frequently.

George P. Bush, then twenty-three and known in the family as

"P.," was a crossover dream: the good-looking, smooth-talking son of Bush's brother Jeb, the Florida governor, and Jeb's wife, Columba, who was Mexican. At an event to honor Hispanic women in Southern California, his uncle introduced him as the real star of the clan. He stood up and blew kisses to the mostly female audience. Afterward, his uncle remarked to reporters, "He's a handsome dude, isn't he?" Some officials in the campaign saw him as a bridge to the youth vote, the Hispanic vote and even the women's vote, which perhaps explained his agreement to be photographed for *People* magazine's most-eligible-bachelors issue in a horizontal, come-hither pose.

In April, Uncle W. was all education, all the time, pushing billions of extra federal dollars every year for reading programs for underprivileged children and recommending that Head Start be put under the auspices of the Department of Education and turned into an early-reading initiative. Education was his centrist trump card, the issue that proved—in a phrase that his aides used promiscuously during this stretch of the campaign, when he was positioning himself for the general election—that he was "a different kind of Republican." He was, but only to a point. What was most conspicuously different about Bush, when it came to other issues, was that he was a craftier salesman than his Republican forebears had been.

That spring, as he rolled out proposals for health insurance and home ownership for lower-middle-income people and talked some more about his tax cut, he framed all of it in terms of a concern for the needy and for social mobility. But he was once again using the stylistics of his presentation to put a progressive spin on programs that were in no way anathema to many conservatives, because these nostrums often included a benefit for the private sector. In suggesting that Americans in certain income groups get tax credits for the purchase of private health- or long-term care insurance, he was also creating a new pool of money for the businesses that sold these policies. In using tax incentives to get house builders to lavish more attention on transitional neighborhoods, he was providing them new sources of revenue. I'm not saying that this effect was his primary motivation, but he paid it no heed when he pitched his plans, which he usually did in settings that placed him smack in the middle of all the hurt, among the struggling masses. His tax cut, he said, was for the single waitress trying to raise two children on her own, which he defined, over and over again, as "the hardest job in America." Never mind that said waitress would only be seeing a few hundred extra dollars, if that. Bush made her the unrivaled object of his compassionate affections.

A reporter covering Bush could bet his or her 401 (k) on certain truths: if Bush attended a Republican fund-raiser in a hotel ballroom packed with rich, fleshy white men, his advertising team and their cameras were nowhere to be found. If he took a walk through a Hispanic neighborhood in Santa Ana, California, there would be a film crew populous and frenetic enough to tackle a David Lean epic. And if Bush stumbled across an aspiring bodega owner who expressed faith in the Republican way, Bush would come to refer to this man in stump speeches as a veritable prophet upon the land.

The campaign's concern over the photographs being taken of Bush and the footage being shot could not be overstated. One night early in the race, Hughes approached Bush to let him know that she was about to let a photographer come snap a few pictures of him, then noticed something in Bush's hand that just wouldn't do. It was a near-beer—a non-alcoholic brew. Bush, who had given up the real stuff back in 1986 and subsequently talked extensively about his sobriety, still liked to drink these, and they came in dark amber or green bottles that made them look like the real thing. Hughes told him that he had to get rid of it, as people might jump to the wrong conclusion, and after a few mild protests from Bush, he finally put his bottle down. Only then did the photo session commence.

This near-flawless public front made any private lapses all the more delicious. There were a few black reporters covering Bush, and on several occasions, when talking to them, he put his foot so deep in his mouth that his toes must have been tickling the top of his stomach. One day, one of them moved a row forward from his usual seat on the plane to talk with me, and when he said something sarcastic to Bush, Bush joked that maybe he should return to the back where he belonged. Even as he said it, he seemed taken aback by the way it could be heard, and the awkward, unfortunate turn of phrase just hung there, like a curious odor. On another day, the same reporter had headphones in his ears and a portable CD player in his lap, and Bush, trying to make conversation, asked him if he was listening to rap. (He wasn't; the disc was John Coltrane.) Bush confused two other black reporters, both television correspondents, calling one of them by the other's first name. They didn't look that much alike.

As thorough as Bush's image makeover was, he rejected any notion that it was taking place, sometimes quite angrily, as was the case when he faced off on the plane one day with one of the television producers on-board. The confrontation revolved around an article of clothing. The Democratic National Committee, in one of its more inspired moments, decided to note Bush's post–South Carolina efforts at restyling by printing up T-shirts that said, "Bob Jones Redemption Tour." The organization mailed these keepsakes to reporters on the Bush beat, and the producer, feeling adventurous, decided to wear his in Bush's presence. Bush was not amused.

He walked merrily down the aisle from the first-class section to coach, laid eyes on it and snapped, "You're off the plane!" Then he marched back to his seat, forgoing his usual conversation with reporters. His threat proved idle, but his pique was very real.

The episode revealed something other than Bush's quickness to anger. It showed that all of us covering Bush knew exactly what he was up to—knew that we were being subjected to a contrived pageant with pragmatic goals. And we indeed reported this: in our stories about where Bush was appearing and whom he was positioning at his side, we included a sense of why he was doing it, of the reparations being made. But such skepticism could not be inserted into the footage and photographs that showed Bush in a Hispanic preschool or a black church, and these pictures were all that many voters, too pressed for time or uninterested to look at the text or listen to the commentary, absorbed. Modern politics wasn't just superficial because the politicians made it so. It was superficial because the voters let it be.

Bush and his handlers paid almost as much attention to language as they did to visual compositions. There were things they wanted to say forthrightly and things they only wanted to insinuate, insinuation being the way to split the difference when it was unclear if a given line of attack would hurt or help. For long stretches of the campaign—stretches that didn't involve destinations like Bob Jones—Bush tried not to tackle Clinton's sexual behavior directly; many Americans thought that too much had already been made of it. So he alluded to it indirectly, a method exemplified by a campaign commercial titled "Pictures." In this spot, Bush said that parents would come up to him at campaign events, show him photographs of their children and say that they wanted their kids to be able to respect the White House. He also incorporated this riff into some of his comments on the stump;

these parents, as Bush told it, never again wanted to see the White House sullied. But Bush was Bush, so the story once or twice received an abbreviated and ludicrous telling, with the parents approaching him, pointing to the photographs and imploring, "Never again!" It sounded like Bush was on the verge of a national sterilization policy.

Although the gaudy theater and emphatic symbolism to which Bush often turned was most apparent as he pivoted from his Republican primary struggle against McCain to his general election battle against Gore, it was evident throughout the rest of his presidential campaign and presidency as well. Those of us covering Bush were most amused by a recurring motif he latched onto in the late summer and fall of 2000, when he would be greeted at airports by what his aides called "tax families," middle-class people whose financial profiles made the case that Bush's proposed tax cut would serve working people better than Gore's. These examples were carefully selected so that they fell into none of the categories for tax savings under the baroque architecture of Gore's plan but fit profitably into Bush's proposals.

Before we met a chosen tax family, Bush's aides would explain in great detail how it would end up saving $1,000 or $2,000 or $3,000 if his tax cut went through but only something like $23.62 if Gore's took effect. The goal was to repudiate any notion that Bush's plan favored the rich. In raw dollar terms, of course, Bush's plan did, and we would joke on the plane about when Bush might serve up, say, the Cheneys as a tax family, with potential savings in the tens or hundreds of thousands of dollars. Needless to say, that never happened: if a family made more than $120,000, its members' chances of standing beside Bush with grateful smiles on their faces was pretty remote.

Bush and his aides so loved tax families that they even staged tax family reunions, or command performances, including one when he finally signed his tax cut into law in the White House in May of 2001. And he recycled the concept by selecting and showcasing what might be called prescription drug families: people who would ostensibly benefit more from his solutions to prescription drug coverage than from Gore's. During the fall of 2000, he traveled with little index-style cards on which he scrawled the pertinent information about these people that he would later recite when the television cameras were all in place and he was standing before them at a microphone. On one such cheat sheet, he had listed the yearly cost of "perscription" drugs.

His yen for this sort of political pageantry traveled with him to Washington, where he turned anew to symbolism to amplify his claim to being a sensitive, compassionate, diversity-conscious and open-minded leader. For an event at the White House at which he announced an initiative to help the disabled, his aides fashioned a seat for him that put him at eye level with people in wheelchairs. His weekly and monthly schedule was like a polyglot, poly-racial, poly-creed fantasia, meetings with Hispanic groups followed by meetings with women's groups, speeches to black pastors followed by chats with archbishops. He went to Catholic University for the dedication of a new center there in honor of Pope John Paul II. He went to the Holocaust Museum.

The fourteen people he appointed in early May to a commission to study the future of Social Security included several African-Americans, several women and even a passel of Democrats: Bush never failed to spot someone on the other side of the aisle who might support his position on a given issue. But all of them were already on board with his push for individual retirement accounts within the entitlement program. A similar dynamic took hold a week later, when he announced his first eleven nominees to federal appeals courts, and administration officials quickly pointed out certain demographics. "The president is putting forward an exceptionally diverse slate of nominees," said Timothy Flanigan, a deputy counsel to the president who helped coordinate the selection process. "Six are women or minorities." This was true: in addition to five white men, the president was nominating three white women, two black men and one Hispanic man. He was even nominating one Clinton holdover. But there was no denying the overall conservatism of the group, many of whom were fierce states' rights advocates or anti-abortion foes or school-prayer proponents or several of the above.

As Bush, with his devout Christianity and daily Bible reading, nibbled at the line between church and state, he called it something else. He framed his push to allow religious groups to receive more federal funds for social work not as an experiment in spiritual healing but as the next offensive in a war on poverty that President Lyndon B. Johnson had begun, semantically allying himself with a liberal Democrat from the 1960s. He cannily framed it as an idea aimed principally at struggling minorities in the inner city and he showcased the black ministers who agreed with his thinking to hammer home that contention.

And when he unveiled his energy plan, the crux of which was more of everything—more nuclear power plants, more oil and gas pipelines, more refineries across the land, more public land pressed into industrial service—he instead emphasized the incentives for energy conservation and efficiency in the proposal. Cheney had previously made clear that conservation was not the administration's foremost concern, giving a speech in Toronto where he all but dismissed conservation as a "personal virtue" with little public impact. But Bush and his aides compensated for that revealing lapse of political judgment by producing books about the administration's energy plan that featured illustrations of beatific wildlife, of trees stretching toward the sky and fly-fishermen wading through unpolluted waters. When Bush hit the road upon the report's release, he brought Environmental Protection Agency Director Christie Whitman, the administration's straitjacketed environmentalist, with him, toted banners that displayed the words "conservation" and "efficiency" and headed for a futuristic power plant in St. Paul, Minnesota, that produced electricity from alfalfa and turkey manure. The gist of Bush's plan had nothing to do with poultry excrement.

Perhaps my favorite bit of packaging came in August of 2001, when Bush took his summer vacation at his ranch. With his previous trips there and his long weekends at Camp David, he had already spent much more time away from the White House than most of his predecessors had, and his aides clearly didn't want him debited for the entire four weeks he planned to retire to Crawford, Texas. They scheduled a few day and overnight trips for him to take from Crawford and, just before he left Washington and headed west, announced that he was not going home simply to rest. He was going "Home to the Heartland," which was defined broadly enough to include a visit to an elementary school in Albuquerque, New Mexico, a state that was conveniently next door to Texas and presumably qualified as the left ventricle. When his spokespeople gave briefings to reporters in a filing center in Crawford, they did so in front of a newly fashioned logo that dubbed Bush's ranch "the Western White House." Message: Bush was working. Which, indeed, he was, but not all day long and not every day.

Then again, it was hard to surpass the packaging during the campaign, when it had surely reached its apotheosis. His aides produced not one but two bound paperback books, with lovely matte covers, that amounted to the collected works of George W.

Bush: the major policy speeches, the position papers, all the greatest hits! Each included a generous and highly selective serving of ellipses-rich quotes from various news sources about the addresses, one of which elicited this rave from Clarence Page, who wrote a column for the *Chicago Tribune*: "If it were a movie, it would deserve an Oscar!"

The campaign was equally invested in its hyper-alliterative slogans, which proliferated like Starbucks franchises. Bush was always a "compassionate conservative," but he started out offering "a fresh start for America" and "prosperity with a purpose," until McCain's reform mantra needed to be appropriated and Bush became "a reformer with results." Hughes told reporters that this was what Bush had been from the beginning of his political career, when he first ran for Texas governor in 1994, but that seemed odd to several of us. How could he have had results when he had not yet occupied any office that would allow him to practice the reforming ways that led to them? Toward the end of the campaign, the Bush operation coined yet another catchphrase, as Bush touted "real plans for real people." We weren't alone in our confusion. At an airport rally in St. Petersburg, Florida, Bush spontaneously rolled several slogans together, telling voters that he would bring them "real results for real people," presumably through reform. Then again, Gore, who had once promised "change that works for working families," switched courses and alternately extolled, from event to event, "progress and prosperity" and "prosperity and progress," never settling on a correct order and never spelling out that these two "p" words surely encompassed a third: purpose. All of which would undoubtedly work to the benefit of working families, or real people, or whatever you dubbed the enormous clump of malleable voters between the poverty line and the Bendel's mailing list.

CHAPTER FIVE

"THE COOLEST THING OF ALL," BUSH SAID WITH a look of dreamy nostalgia, "was to light up a butt." He was talking about cigarettes, which he had smoked as a young man, at least before he moved on to Copenhagen snuff tobacco and, later, Beechnut long-leaf tobacco. With those quasi-addictions, he reached a point where he no longer had to spit the juices out; he just swallowed them, his body craving and loving the liquid nicotine. This was the state of things in the early 1970s, when he attended the graduate business school at Harvard University and adhered to a pre-class ritual that helped carry him through even the most tedious lectures. "I used to walk into class with a big cup of black coffee and a chew in," he recalled, the recipe for an electric buzz. Summoning the memory of that feeling and those days, Bush smiled and shook his head.

We were on our way to the Kentucky Derby, on the first Saturday of May, and the campaign plane was mostly empty—this was a dubiously newsworthy event during a pronounced lull in the presidential race. Bush hovered over those of us seated in the middle of the coach section of the plane and reminisced. Maybe it was the sparseness of the press corps that got him going and goaded him toward such candor. Maybe it was the fact that the Derby was the kind of party he would have enjoyed in a much different, more expansive fashion when he was younger, before he renounced his worst habits and grew more serious about everything in his life. "You should have seen me about twenty years ago," he told us. "I would have been betting and drinking all at the same time."

He rolled a verbal reel of the highlights of his youth, of some of the things he remembered most fondly and best. At Yale, he said, he drank too much, a habit that followed him for the next two decades and increasingly became a drag on everything else he was trying to do. His daily runs were not just about fitness; they were a purging ritual, a way to shake off the aftershocks of overindulgence. He suggested, with a reference to the way alcohol "competes with your affections, with your family," that it made him a worse husband, a worse father. He acknowledged that there existed, in theory, some midpoint between the amount he would consume at a social get-together at night and his total and utter abstinence from alcohol later in life, but, he said, "I couldn't seize it." Once he stopped drinking, he added, "I discovered that you don't have to be blotto to be self-centered and talk openly with people." That was how he often defused serious subjects like this—with a self-effacing dig.

He had deep scratches and some bandaging on one of his hands. The culprit was Ernie, the six-toed cat who had appeared at the governor's mansion one day years ago and thus found a rather exalted home, where the Bushes named him for Ernest

Hemingway, onetime owner of a pet with a similarly freakish digital endowment. We knew this because nothing set Bush to talking like questions about his pets. He constantly regaled us with stories about them, and this one involved carrying Ernie outside, where a sudden screech caused the cat to dig his claws into his owner. Bush bore Ernie no ill will. "I was *compassionate* with him," Bush said, mocking one of his campaign slogans. "He's a great man."

The Derby was a day-long photo opportunity, the epitome of political vapidity, allowing Bush to bless a hallowed tradition in a state that, at least then, seemed likely to shape up as one of the important battlegrounds between him and Gore. (It didn't; Bush pulled ahead early in the polls in Kentucky and never lost his lead there.) Bush stood beside the thoroughbreds and admired them, wrapped his arms around the local politicians and mingled with many famous people he expected to encounter and one or two he did not. Out of nowhere, the actress Bo Derek appeared, in an enormous hat and an orange shirt and white slacks that were tight as a second skin. Bush smiled gamely, seemingly unsure if her presence was a political plus or minus, and she lingered near him as he stood in a dirt area beside the track to greet the winner of the fifth race. It turned out to be a horse named "Don't Be Silly," a fitting instruction for Bush, and the jockey astride it was Hispanic, a fitting detail for the diversity-conscious candidate. The only things in a political campaign that came faster and easier than food were accidental metaphors.

On the way back to Austin, I fell asleep, a baseball cap pulled low over my sweat-soaked hair, sunburned forehead and the tops of my bourbon-swollen cheeks, then awoke to laughter and a slight itch near my right ear. Bush had knelt down in the aisle and pressed one side of his face against mine so that a television producer could turn it into some frivolous footage. He was in mischievous spirits. He returned to memory lane, remembering that he once took Latin but saying he could now recall only one phrase, "ubi ubi sub ubi," a pun that he translated as "Where, oh where, is your underwear?" He looked forward to having the next day, a Sunday, off. What would he do? Go to church, he said, and then relax. "I'll be reading policy," he predicted with a deadpan expression. Then he grinned. "No, I'll probably take a few naps."

The stretch of the campaign from late March through early July, when the primaries had receded into memory and the political conventions and the start of the general election were still a ways off, was unlike any other—slower, less hurried, less taut. Most news organizations had pulled their emissaries from the plane, and the press corps had dwindled to the faithful few, representing the biggest-spending news organizations and the Texas publications for which Bush was not only a national story but also a local one. On a day like the Derby, even many of these news outlets begged off. The atmosphere on the plane grew unusually relaxed, and Bush spent longer spans of time among his shadows, trying to badger us, trying to seduce us, strutting for us, confiding in us.

It was as if he had been waiting for months to exhale, and the breath rushed out of him in a sustained, gale-force sigh of relief. His conversations with us made it clearer than ever that he was most content to talk about anything other than politics, anything other than policy. They also reaffirmed that the man still leading in all the major polls was someone whose tolerance for the task before him had finite bounds. Left to his own devices, Bush discussed the flora and fauna on his ranch, dwelling on one or two occasions on a Red Russian stag that had apparently escaped from a wildlife preserve and wandered onto his property. Bush professed a particular fascination with its attempts to mount the cows and expressed a special concern with finding it a zoologically appropriate mate. Bush, it turned out, had a bit of a bawdy streak, and quizzed aides about the social lives of reporters, indicating an interest in who might be coupling. From the reports he received, he often knew who had been in the bar latest on a given night.

He prattled on about baseball, hunting, baseball, running, baseball, lifting weights, baseball, escapades from his youth, baseball and good times with his family. Did I mention baseball? Bush had once, long ago, dreamed of a career in the sport and, much later, paved his path to political office by buying a piece of the Texas Rangers, which dramatically raised his profile in the state. Now, he coaxed reporters into asking him to recite the most arcane baseball trivia, inviting the kinds of pop quizzes he resisted on any other topic.

In ways small and large, he reminded us of how strangely the august position he was seeking squared with the colloquial person he was and wanted to be. His reveries on the way to and from the Derby underscored how long in life it had taken him to get his act together. His reveries on other flights and in other settings illuminated how unusual this act remained—how narrow the parameters of his field of interest could be, how fitfully he tried to expand them, how self-consciously he tried not to seem too serious or too ambitious. He had once, months earlier, told an audience in Iowa that he not only didn't want to be president when he was growing up but also didn't want to be president "until recently." He now told us, more or less, that he wasn't claiming to be the perfect person for the job; he was claiming to be the best of the limited choices before voters in the 2000 election.

"There are a lot of people who would make great presidents," he said on a flight from California back to Texas in May. His choice of words as he continued was intriguing. "And the problem," he added, "is none of them choose to run. There's just two of us."

The second person, of course, was Gore, the presumptive Democratic nominee, who was certainly more experienced and arguably more interested, for whatever this was worth. But Bush had the good fortune of Gore's own shortcomings as a candidate, a set of problems wholly different from Bush's, and the contrast between the two men's bearings during the long, quiet buildup to the conventions could not have been more stark.

This was not, by a long shot, the most exciting chapter of the campaign; it was filled with more than the usual share of pointless photo opportunities, like Bush's participation in a Fourth of July parade in Belton, Texas, a fleaspeck of a burg about 55 miles north of Austin. For this event, the dozen or so media stalwarts who showed up were loaded into a sort of pen on the back of a flatbed truck that trundled along ahead of Bush—who was walking—at an almost undetectable speed of perhaps 1.734 miles per hour. To pass the interminable hour during which Bush paraded under a Texas sun that resembled nothing so much as a flamethrower with a reach ending two inches above our heads, we basted ourselves in sunscreen, counted the sweat droplets on our notepads, watched the dark circles under Bush's arms colonize other areas of his shirt and tried to divine a pattern in his movements from the center of the road to the scrums of people flanking it.

"He's going to the left," ABC's Berman would say, as Bush shook hands with people on that side.

"He's back to the right," the *Los Angeles Times*'s Miller would announce.

"I'm counting more trips to the right, but plenty to the left," I would decide. "It's pretty even."

"That's because he's a centrist," Berman offered.

"It's a metaphor," we'd all agree, and then one of us would pass the sunscreen to another, whose last layer had melted so thoroughly it was now sliding down his ankles, into his sneakers.

On the plane, the rules of engagement were initially awkward—endlessly negotiated and never quite satisfying to any of the parties involved. Political campaigns can be messy that way. Reporters want access, because it breeds insight, but they don't want to give up anything in return. Candidates want to please them, but also want guarantees—off-the-record agreements, the right to take back an unplanned, imprudent revelation. Bush and his aides at first demanded that his extended sessions of banter with us be unprintable and unusable, and for a short while in late March we abided by that, because we did not see how the alternative served us any better. If we said no, Bush would stick to his seat in the front, and we would be sacrificing the ability to get a sharper sense of him, which was bound to inform our coverage even if it could not, by verbal contract, yield specific information and verbatim quotes.

But the more he came back, the more we felt exploited. He was obviously using—and relishing—this opportunity to try to charm us, to get us to see him as flesh and blood, to personalize the situation to the point where we might think longer and harder about reporting anything derogatory. And yet, with each passing day, we were learning less and less that was genuinely revelatory about him; we were just listening to, and then being forced to ignore, quotes and anecdotes we wanted to use. McCain had operated without such a safety net. So a few of us rebelled, and finally announced to Hughes and her assistants that while we welcomed Bush's presence, we were not going to grant him any protections, and he should know this and do with it what he would. We did not see him for the next flight or two and then, lo and behold, he returned. We extracted our notebooks and tape recorders, which had the effect of damming his words, so we put them away. We had no qualms about easing him into this, and we could always take out these notebooks again if he said anything important, or we could scribble general recollections of the conversation after he left. Besides, we learned that if we treated these encounters as formal interviews, it was pointless: he retreated into precisely the same language and answers he was giving at the news conferences that he regularly held on the ground, or he shut up.

I adapted by keeping my laptop open and typing furiously, my eyes affixed to the screen. He and most everyone else around me assumed that my committed efforts meant I was writing a story, but what I was really doing was putting down as many of Bush's comments as I could. I didn't have to be looking at him to do that, and as long as I didn't, he seemed not to suspect me. I soon detected that a few other reporters had happened upon the same strategy, and we all gradually began to pool our notes, fleshing them out, and to put bits and pieces of Bush's remarks into feature stories about his personality and mood. He still came back. Before doing so, he would occasionally murmur to an aide up in the front with him that it was time to "feed the beast," an indication that there was a method to his ministrations and a sign of his essential distrust of the press. But the full truth was that he also enjoyed what he was doing. Like many politicians, he was a performer, and his excursions to that midsection of the plane provided him with a mostly rapt audience.

They also passed the time. Much earlier in the campaign, when he was ending a voluble saunter up and down the aisle, he groaned, rolled his eyes and said he was compelled to return to the front, because a dread "policy briefing" awaited him there. It was, for him, the less attractive obligation. Much later—when the plane filled up again, Election Day loomed closer, every word he uttered drew more scrutiny and he decided to end his visits to the back—he would stand at the invisible barrier between first class and coach, look longingly and dejectedly at some of us and shout little greetings and jokes. He declared that as long as he stayed on his side of the line of demarcation, his remarks were off-limits, although many of us, never having agreed to that, paid the wishful restriction no heed.

He once placed his foot just over the boundary, teasing us. "On the record," he grinned.

He pulled his foot back to the other side and banished his smile. "Off the record," he announced.

Another step forward: "On the record."

Another step back: "Off the record."

And then he laughed like a little boy. He was clearly caught and torn between where he wanted to be and where he needed to be, between what was fun and what was prudent.

Only once during the spring of 2000 did he consent to a full-blown, semiformal news conference in the sky, and there was no question about what prompted it. Ceci Connolly, a *Washington Post* reporter who usually covered Gore's campaign, had ventured into Bush World for a few days, around the time that Bush was flaunting his celestial accessibility and his terrestrial habit of giving reporters about fifteen minutes three times a week to ask him about policy or politics. Gore, by contrast, had gone for two months without indulging such questioning in the air, on the land or, for that matter, in the water. The Republican National Committee, on its Web site, maintained a count of the number of days the vice president had held reporters hostage, replete with humorous illustrations of well-known journalists in stockades. Bush was showing Connolly just how much more hospitable his operation was. She turned on a tape recorder and Bush didn't go away, so we all turned on ours. The give-and-take lasted well over an hour.

What was most remarkable about it, and was probably also triggered by Connolly's presence, was the venom he vented toward Gore. Gore had been on the attack lately, blasting Bush's proposal for a partial privatization of Social Security, the state of the Texas budget and Bush's record on crime, and Bush deemed all of these critiques to be gross misrepresentations of his positions and record. "Someone running for the highest office of the land should stick to the facts," he said, a seemingly throwaway line that actually foreshadowed a strategy most conspicuous after a debate later in the year, when the Bush campaign successfully mined Gore's answers for exaggerations and essentially branded him a liar. "Pretty soon," Bush continued, prophetically, "it's going to have a corrosive effect on his campaign if he's not telling the truth."

Bush was bound for California, where he would attend a Cinco de Mayo celebration, and a reporter noted that Gore, whose first grandchild had been born on the Fourth of July, sometimes told Hispanic audiences that he hoped his second grandchild would be born on Cinco de Mayo. Bush looked nauseated, turned to one of his press aides, Gordon Johndroe, and muttered, in a voice wholly audible to several of us, "Totally pathetic."

"What I'm not willing to do," he told us, "is sell my soul to become the president." Was he saying that his Democratic opponent was? "That's for you all to figure out," he answered, leaving no doubt about the conclusion he had reached.

Few of his conversations with us during this springtime stretch of his campaign were as politically pointed, but some had other purposes or provided equally noteworthy glimpses of Bush. He made a number of cutting, cunning observations that suggested an ample if sometimes undernourished intelligence. On the flight after a satellite hook-up with a convention of newspaper publishers, during which the highest-ranking executive at the *New York Times,* Arthur Sulzberger Jr., had asked him a first question, Bush made his usual journey to the back of the plane. He saw me, stopped a few feet away, looked to the side as a bit of inspiration struck him, then smiled and said nothing for a few seconds, creating a silent drum roll before his remark.

"Do you think," he asked, "that Sulzberger worked his way to the top?"

Sulzberger belonged to the family that had owned the newspaper for generations. And with this little question, Bush made a big point: that he was not the only one in the world to benefit from his background, and that the advantages that people like he and Sulzberger enjoyed did not mean they were undeserving of their stations and responsibilities. Bush was not looking for an answer. He nodded—as if to say, "Get it?"—and then walked away.

During another flight, Bush proudly told several of us, "I don't read half of what you write."

"We don't listen to half of what you say," shot back a wire-service reporter, thinking he had found a pretty fitting last word.

But Bush stole it from him, saying that this habit of reporters was abundantly apparent in the half of their coverage that he indeed read.

Bush liked to claim that he was on a "news media blackout" when it came to coverage of the presidential race, because he didn't want to be upset by what he saw or hold grudges against correspondents and producers that he had to see every day. It wasn't entirely true. He occasionally complained to a reporter from a given news outlet about something another reporter had written or said, blowing his cover. He then tried to insist that Hughes or Laura Bush had brought the offense to his attention, but in some cases the specificity of his recall made this seem impossible—or made clear that once he had been alerted to the crime, he dove in and investigated it himself.

In any case, it was a peculiar boast: he was taking pride in being willfully uninformed. But it dovetailed perfectly with much of his behavior during the spring, as he mingled more extensively with reporters and more readily indulged his I'm-just-a-regular-and-fun-loving-guy whims. One day, he walked up and down the aisle of the plane with Billy Bass, one of those battery-operated rubber fishes that can be mounted on the wall and made, with the press of a hidden button, to belt out tunes, some with aquatic references. This one did renditions of "Don't Worry, Be Happy" and the Talking Heads classic "Take Me to the River," and Bush prodded it into song again and again, making sure each and every person in his path—reporters, campaign aides, even Secret Service agents—had taken proper note of the trick, which amused him to no end. Later on, a Texas reporter asked one of Bush's aides where Bush had got the toy, and the aide responded that it was a gift from "a friend." The aide later confided that the reason for the vagueness in that answer was that Bush had in fact received dozens of Billy Basses from acquaintances wide and far. His penchant for broad, simple humor was that well known.

Before an outdoor speech in the Southern California desert, Bush pledged to keep his remarks brief, in deference to one of the heat's less attractive effects on the human body. He tugged on the portion of his dress shirt covering his underarms and explained, "Pit City," a place we would later come to gaze upon during that parade. He glanced in horror at the slivers of sushi that we had been served during one flight and held his peanut-butter-and-jelly sandwich high, like a chalice. "This is heaven, right here," he proclaimed. Much of the time, he greedily consumed bilious orange Fritos or Cheez Doodles, stacked high in clear plastic cups, and paid verbal homage to the joys of hamburgers and tacos. When he ate French fries, he dipped them into puddles of ketchup deeper and broader than anyone over the age of twelve typically amasses.

He once treated a few reporters to his cooking secrets for wild dove, which he liked to hunt when it was in season. "Well," Bush said, "you gotta pluck 'em, breast 'em, put 'em on the grill with maybe a little bacon and jalapeno . . . you can marinate them. I'm probably just going to toss them right on the grill. Baste them, as we say. I'm not the world's best cook, to be frank with you. Let me just say this: They're not going to be raw by the time I get through with them."

These were the eating habits and culinary musings that complemented his references on the stump to undecided voters as "tire kickers" and his promises to get to the point "right quick," a lingo so rife with Southern spices, Texas twang, homespun truisms and suspicion of city slickers that a listener could almost picture Bush riding up to his ranch in a red pickup, a mound of fresh kill for dinner in the back.

When he talked on the plane about a formal party he was obliged to attend, he winced at the thought of wearing a tuxedo and vowed not to be photographed in one. It was unacceptably pompous attire. He gloated over his ability to withstand the summer temperatures at his ranch, which were not unknown to inch up to 105 or 110 degrees, and over the likelihood that it would wilt and sour the moods of reporters who stayed nearby to keep a body watch on him. He chortled that it was nothing like Kennebunkport, Maine, where his father had traveled in the summer and the press that followed him had been all too happy to stay. The son, with his cowboy boots and cowboy swagger, was going to rough it on the scorched earth of central Texas. Sissies needn't apply.

He made fun of people too invested in learning, which he saw as a poor substitute for horse sense, and railed against the Northeastern elite. When he mentioned one day that one of his fraternal-twin daughters, Barbara, was applying to Harvard, his lip curled and he seemed to choke on the syllables, as if they added up to the dirtiest of words. (She ultimately decided on Yale, which made him happier but was hardly any different.) He gave ABC's Berman, a Harvard alumnus who rarely mentioned this fact, an advertisement, torn from an airline magazine, for audio tapes that would provide the interested listener with "a Harvard graduate's vocabulary in just 15 minutes a day!"

"I have just ordered several," Bush had scribbled across the page. "What do you think? Do I have a chance?"

His political allies took their cue from this shtick, rendering the oddest and least flattering testimonials that modern politics had ever witnessed. At a rally in Albuquerque on the last day of May, New Mexico governor Gary E. Johnson recounted for hundreds of voters a conversation that he and Bush had once had about the speakers at a conference of state leaders.

"George turns to me," Johnson recalled, "and says, 'What are they talking about?' I said, 'I don't know.' He said, 'You don't know a thing, do you?' And I said, 'Not one thing.' He said, 'Neither do I.' And we kind of high-fived." Johnson went on to salute Bush as that rare "somebody who will rather admit—or rather talk about—the things they don't know and make that really evident." It was high praise indeed, but no higher than the encomium awarded by J. C. Watts at a rally in South Carolina months earlier.

Watts had told voters that a president needed sturdy values, not a superior intellect.

"You can buy clever," Watts said.

This aggressively flaunted disdain for intellectualism and anything that smacked of excessive formality represented an overreaction to Bush's breeding, to the whiff of blue blood that had prodded journalists into deeming his father a wimp and questioning the elder Bush's capacity to connect with average Americans. Bush liked to cast himself as a son of Midland, Texas, and Midland alone, the dusty bowl in which he said he wanted his ashes scattered someday. But the truth, of course, was that he had also been reared in the richer enclaves around Houston and, come summertime, along the coast of Maine. The truth was Andover, Yale and, yes, Harvard. The truth was a paternal grandfather, Prescott Bush, who had represented Connecticut in the U.S. Senate and taken him to places where there were finger bowls on the table and young George knew that a friend of his should not be sipping from one of them. This was also who he was, and he had spent a lifetime compensating for it, rebelling against it and calculating that it served him best to separate himself from it.

Much of his distaste of haughtiness and pretension, especially liberal haughtiness and pretension, was surely real. Bush had long cited as one of his most influential experiences the time a chaplain at Yale approached him and reacted to his father's defeat in a closely watched Texas Senate race by remarking that he knew both contenders and that the elder Bush had lost to "a better man." For decades to come, the younger Bush nursed his anger and maintained a studious distance from the university. At Harvard, he kept to himself rather than listen to liberals lambaste Republicans in the midst of revelations about Watergate, and he came to deplore what he saw as their preachy certainty. He was never going to be like this, never going to behave like he was better or more knowledgeable than anybody else. It was like a religion to him.

But it also made political sense, a lesson he learned when he ran for Congress in the Midland area in 1978. He had returned to West Texas, his former home for more than a decade during his childhood, to do what his father had once done there: make his way in the oil business. He saw himself as a man of these people. But during that 1978 campaign, which he lost, his Democratic opponent, Kent Hance, portrayed him as the opposite of that—as an elitist interloper. At the time, Hance publicly told a joke about standing on a rural road when a fancy car came along. Bush was the unacknowledged driver in this tale.

"It was a Mercedes," Hance said. "The guy rolled down the window and wanted to know how to get to a certain ranch." So Hance gave him directions, telling him to turn right after a cattle guard, which was the name for a kind of metal grate that kept livestock from straying. "He said, 'What color uniform will that cattle guard be wearing?'" The finishing touch on the joke, as Hance rendered it, was that the Mercedes had Connecticut license plates.

Hance also made fun of a campaign commercial that showed Bush jogging on a track, an uncommon activity for that area at that time. "The only time folks around here go running," Hance observed, "is when somebody's chasing 'em." Decades later, reflecting back on that campaign and on the kind of political identity Bush would later forge, Hance said that Bush clearly decided he wasn't going to let himself be "out good-old-boyed again. He's going to be the good old boy next door."

This is who and where he was in the 2000 presidential race,

except for when he wasn't. Bush wanted to be more folksy and funny than Gore; he wanted a much different political identity than his dauphin doppelgänger had. But the altitude of presidential politics demanded that he not let himself be cast as an unreconstructed rube and that he bridge what might be called the gravitas gap between him and Gore. So he searched for a vague, happy middle ground that took into account the ambivalence of American voters, their appetite for candidates who fell somewhere between uncommon erudition and common knowledge, who were at once ordinary enough to understand them and extraordinary enough to lead them. He tried, in other words, to have it both ways—a balancing act that would later be on especially vivid display in the weeks after the World Trade Center fell, when he seemed in some ways to have finally perfected it. One of the most interesting tensions in his campaign was his and his advisers' simultaneous assertions that book learning and highfalutin logic weren't important and that Bush was much better educated and more thoughtful than the "punditry," another dirty word in his personal dictionary, gave him credit for being.

In May, during an interview with the journalist Chris Matthews for the television news program *Hardball,* Bush remembered that he had once publicly told Matthews that his favorite author was Churchill and that a commentator had subsequently deemed it a planned effort to please Matthews, a known Churchill devotee. This miffed Bush. "Do you think," he asked Matthews, "that I'd take time out of my life to research what the hell you like?"

A few days later, one of Bush's aides flashed a panic-stricken expression when Bush strode into the coach section of the plane cradling a book titled *Spanish for Dummies.* It was not Bush's, and he was merely using it to tease a reporter, but the vigilant, nervous aide looked this way and that, obviously worried that someone with a camera might snap a picture of Bush with the word "dummies" right below his chin. This would apparently be taking the good-old-boy-next-door image too far.

Months later, Bush had an exchange with several of us on his plane that demonstrated his eagerness to get points for the American history he had retained. Candy Crowley, the principal campaign correspondent for CNN, jokingly asked Bush if she could leave the plane through the less crowded front door, which he used, instead of the back door, noting that the rear exit was "separate" but not "equal."

"Plessy versus Ferguson!" Bush exclaimed, like a proud student, showing off his familiarity with the 1896 Supreme Court decision that instituted a separate-but-equal doctrine to permit racial segregation in public facilities.

Bush readily shared his reading list with reporters. During one of my initial interviews with him, in October of 1999, as he flew in a small charter from North Dakota to Texas, he proudly brandished a copy of *Titan*, a 774-page biography of John D. Rockefeller, and pronounced himself a serious history buff. He fingered the ribbon-like marker that tracked his progress, which not only drew attention to how deep into the tome he had traveled but also allowed him to explain that it was a souvenir from the annual book fair in Austin. He then volunteered that the book fair was one of his favorite events. He let us know that he was reading *Balkan Ghosts*, about that region. He let us know that he was reading *Lenin's Tomb*, about Russia, and *A Great Wall*, about China. Beyond these titles was an infrequently noted fact: Bush was, in fact, a pretty steady consumer of books. I had erred in the article that discounted that possibility.

Condoleezza Rice, who had become the campaign's chief

foreign-policy adviser, was recommending most of what was on this list, and I ribbed Bush about how monochromatic the collection was, how unvaryingly dutiful. Didn't he ever read fiction? He said he did and would, and asked me to recommend something, especially if it were in the detective genre. I told him I'd do better and bring him one of my favorite detective yarns, *The Concrete Blonde,* by Michael Connelly. Little more than a week after I gave it to him, he said that he had finished it, and provided a mixed review that suggested he had indeed read every word. A week or so after that, he handed me a semi-literary novel that he had also recently finished, *In the Lake of the Woods,* by Tim O'Brien, about a rising politician whose past comes back to undo him. Toward the end of the campaign, he sent his personal assistant back to my seat with yet another book, a mystery by Ross MacDonald titled *The Zebra-Striped Hearse.* It revolved, in part, around the no-good, rebellious child of a rich, accomplished man. It was also terrific. Bush had perfectly decent—and psychologically fascinating—taste.

And yet, despite these indications of a moderately active mind, Bush betrayed gaping holes in his knowledge during his conversations with us. Whether he was reading campaign articles or not, his attention to other fare in newspapers and magazines was extremely scattershot; we constantly made references to recently reported facts or details that he had somehow missed.

Somebody asked Bush to respond to a criticism from Joe Andrew, and Bush could not place the name, even though Andrew was the national chairman of the Democratic National Committee. Somebody asked Bush if he thought that Angel Maturino Resendiz should be tried outside of Texas, and Bush said, "I don't know that issue. I'm sorry. I need to look into that." Resendiz was a serial killer whose crimes and arrest had been reported extensively on the front pages of Texas newspapers and in Texas newscasts, and when Bush was reminded of this, he asked, "Isn't he going to be tried in Houston?" Resendiz had been, the previous month, and had been sentenced to death, but then the waiting list for lethal injections in Texas was fairly long.

Most stunning of all was his estrangement from nearly all aspects of popular culture. Reporters on the plane were constantly shifting their eating habits, trying to stay a few paces ahead of obesity, and when someone mentioned the word "vegan" one day, Bush looked confused.

He didn't know what that was.

Bush had sought to play matchmaker between two reporters edging up to an ill-advised involvement and, when it didn't pan out, cracked that the woman in question had shed ten or more pounds "for nothing." But when somebody suggested he was a bit of a "yenta," he flashed befuddlement.

He didn't know what that was, either.

When it came to movies and music and best-selling fiction, the man had apparently never picked up a *People* magazine and never surfed the channels and rode the wave of *Access Hollywood* or *Entertainment Tonight,* not even for a few guilty minutes. He asked NBC's Pelosi if she was going to use the footage from her handheld camera to make a documentary in the style of Roger Moore. He meant Michael Moore. He had to be told who Stone Philips was, even though *Dateline,* the news program that Philips hosts, was an inescapable ingredient of just about every night's prime-time lineup on NBC. He had to be told who Leonardo DiCaprio was, even though *Titanic,* which Bush had not seen, was the highest-grossing movie ever and DiCaprio was idolized by teenage girls only a year or two younger than his daughters.

In a May interview with *Glamour* magazine during which he

gamely submitted to a "verbal Rorschach" experiment—he was supposed to give quick responses to cultural cues—his face "blistered in a purple fury," according to the writer, David France, when France said, "*Sex and the City,*" Bush apparently thought it was an inquiry into his erotic and geographic whereabouts. "Governor," explained his aide Gordon Johndroe, trying to dispel the confusion, "it's an HBO television show." Bush had never heard of it.

In an interview months later with the ABC news program *Good Morning America,* the journalist Diane Sawyer told Bush that Senator Joseph Lieberman of Connecticut, the Democratic vice presidential candidate, "thinks that *Friends* is too racy to be running at the early hours" and "should be later in the evening."

"Do you think it's too racy?" Sawyer asked Bush.

"The movie *Friends*?" Bush replied, not only oblivious to the hit comedy's existence but obtuse to the very wording of Sawyer's question, which had signaled that *Friends* was a television show.

Talking about some of the authors who had been featured at an annual literary conference that his mother organized, he mentioned John "Gresh-um." He not only mispronounced the popular novelist's name resoundingly enough that it could not be attributed to his Texas accent, but also characterized Grisham's work in a fashion that indicated either utter ignorance of his novels or an extremely low tolerance for even a few adjectives. Bush said that he believed he had once started something Grisham had written, but had been put off by the pages upon pages of dense description.

At long last, the Republican party had nominated its first baby boomer for the presidency, and the man they had chosen was no more culturally "with it" than Bob Dole, the septuagenarian previous nominee, had been. In college, during the late 1960s, Bush had listened more often to Motown than to the Beatles. The extent of his commentary on modern theater was a professed affection for the musical *Cats*. His idea of cinematic perfection was *Cool Hand Luke*. He once named Chuck Norris, the martial artist, his favorite movie star. The last time he had fallen head over heels for a screen siren was when he saw Julie Christie in *Doctor Zhivago*. Except for his stated fondness for Van Morrison, he was artistically atavistic, someone who could just as easily have gone to high school and college in the 1950s as in the 1960s.

What made it all the more striking were Gore's ceaseless exertions to prove his up-to-the-minute tastes. Gore had not only seen and, by his own account, loved the movie *Being John Malkovich*, but also hired its director, Spike Jonze, to make the brief biographical documentary about him to be played at the Democratic National Convention. It was a self-conscious assertion of his own modish sensibility, and it did not, in the end, serve him all that well. Jonze included in the documentary the fact that Gore, at home, would rewind videotapes and start them over again if family members got up, wandered off to the kitchen or bathroom and missed a scene or two or even the opening credits. Oh, yeah, *that* was cool.

I pulled a reverse Connolly, dropped in on the Gore campaign and saw that as far as Bush had pushed himself to the silly pole of the behavioral spectrum, Gore had pushed himself even farther to the stiff, serious end. The press section on Air Force Two, the plane in which the vice president flew, was decidedly less festive, and Gore rarely visited it. There was little chit-chat with reporters and almost no casual contact, although Gore's volubility on the stump put Bush's meager offerings to shame, leaving little doubt that Gore was better acquainted with the policies and issues he was discussing—nay, *pontificating* about. This was Gore's curse: He made everything sound like a university seminar, signaling his transition from point to point by saying, "Now," exactly the way a professor might. His follow-along-with-me-if-you-can tone of voice followed him everywhere he went, even into posh private homes where he addressed small clutches of big donors over cocktails or dinner. At one of these gatherings, in a penthouse along the Potomac, Gore's dozen or so pupils stood at the edge of a plush living room, wine glasses in their hands, as he submitted them to a mini-tutorial on the evolution of the economy from the dark valley of stagflation in the late 1970s to the current peak of prosperity.

Bush did not allow reporters to attend such small fund-raisers. Gore would let one or two reporters in, but only to hear whatever remarks he delivered, and those reporters were responsible for a "pool" report that told their colleagues everything they had witnessed. I had volunteered for pool duty that night at the penthouse, because I wanted to make the most of my abbreviated stint with Gore, and I sat alone in a den, waiting to be allowed into the living room for Gore's remarks. I had not met him yet, but suddenly he was walking toward me, because I was along the path to a bathroom he wanted to use. I quickly stood to introduce myself. He coldly shook my hand and, in a voice no warmer than his grip, said, "I know who you are." Judging from his demeanor, this was not such a good thing.

I don't know if the statement was prompted by a belief, widespread in Democratic circles, that the reporters assigned to Bush were letting him off easy, or if it traced back to an article I had written about Gore's wife, Tipper, about a year earlier. It had been a rare unflattering portrait of her, but it had been true to my off-kilter experience. Her staff had kindly obliged a last-minute request for an interview with her to coincide with a mental-health conference she was leading the following week, but when I got to her office, her press aide had told me I would get just fifteen minutes with Tipper. The aide also told me that I needn't bother trying to negotiate for more, because Tipper was due elsewhere and would have to stop the interview precisely at the predetermined time. When the interview began nearly ten minutes late, I assumed I was down to five or six minutes, so I jumped right in, only to have a startled Tipper ask me why we couldn't take a few moments to get acquainted and talk about the story. Her aide then instructed me to do so. The precious seconds ticked by.

I grew panicked and Tipper grew edgy, responding to many of my questions in a curt, defensive manner. When I asked if it had felt awkward to have to consider her husband's political career before her recent public admission of her struggles with clinical depression, she heard the question as an accusation that she was exploiting her illness for political purposes. "You don't understand at all," she reprimanded me. Her aide interrupted several times to clarify or amend something that Tipper had just said. I wrote it as it was, with the dose of attitude I thought necessary to capture the flavor of the encounter, and for many months afterward, reporters who knew Gore's closest associates told me that my name was mud in those circles.

Even so, it seemed self-destructive and counterproductive for Gore not to put on a friendly face. He made no effort. His energies were channeled into his campaign trail remarks, so dense with knowledge, so showy with digressions. He sweated the big stuff and muffed the small stuff. About two weeks after I was with him, my friend Wilson of Reuters left the Bush campaign to drop in on Gore and happened to watch the vice president participate in a kickball game with children at a YMCA summer camp in the St. Louis area. She told me that at first, whenever a kid kicked the ball in Gore's direction, he scooped it up and threw the kid out at first base. House Minority Leader Richard Gephardt, a Missouri

Democrat, then whispered something to Gore, and Gore started going easier on the kids.

Bush, meanwhile, forged ahead with his joke-telling, ever-present effervescence. If he crossed paths with a physician at one of the innumerable panel discussions that his campaign arranged, he might well call the person "Doc." He greeted Wilson, a native of rural Australia, by musing: "Outback woman! One of the things I'd like to do sometime is have dinner with you at the Outback Steakhouse. That way, we'd have the Outback woman at the Outback Steakhouse." If Gore was guilty of a seriousness that was intermittently oppressive, Bush exhibited a cheerfulness that was occasionally fatuous, and neither extreme was optimal. In order to win, and in order to lead a country through anything difficult that might confront it, Bush was going to have to budge some.

CHAPTER SIX

THE BUSH FAMILY'S KENNEBUNKPORT COMpound occupies a bent finger of land jutting out from the Maine coast, and it includes a saltwater pool with an adjacent sauna, a tennis court, a dock, three guest cottages, another cottage devoted to office space and then the big house itself, with picture windows that take in the dusky blue breadth of the Atlantic. It's a private peninsula, identified on maps as Walkers Point, named for the former president's forebears, owned by the family for about a century and easily mistaken for a Club Med that floated north and shed its palm trees and acquired some breeding along the way. The Bushes actually use electric golf carts to get from one area of the property to another, and it was in one of these that the forty-first president of the United States pulled up to greet me late one morning in early July. Taking in the scene as I watched him approach, I was reminded that George W. Bush's extraordinary luck had been more than just a recent phenomenon. It had begun the moment sperm hit egg.

I had seen the elder Bush and his wife, Barbara, from a distance during the campaign, and he had given me written answers to questions I had e-mailed his personal assistant while the former first lady had talked with me once on the phone. But I had never met them, despite repeated requests for face-to-face interviews. When they finally consented, it was a relatively rare exception to their general policy of public distance from the race; for about a year, they had not sat together to talk at length with a newspaper or magazine reporter about their eldest son's presidential prospects. I was to be given one hour with them, no more and maybe even less, and I was frantic with worry that they would speak in gauzy, innocuous, risk-free sentences until the clock ran out, leaving me with nothing to show for the encounter.

The day before the interview, I arrived in Kennebunkport ridiculously early. I had built in enough time for the two legs of my flight from Austin to Portland, Maine, to have mechanical problems, weather problems, crew problems—even an unspecified case of aerial ennui—and *still* get me there before midnight, so I could sleep a solid sleep. I had built in enough time for the possibility of getting so lost on the road between Portland and Kennebunkport that I ended up in a different time zone and was forced to circle back. When none of that happened, and I found myself at my destination in the early afternoon, I took a run—well, a brisk, aerobic waddle—from my waterfront lodge to the main gate of the Bush compound. That way, I would know the *exact* route and distance to my appointment, scheduled for 11:30 A.M. the following day. I set my alarm for something absurd like 7:00 A.M., just to make sure I had ample opportunity to review some of the questions I wanted to ask and some of the clips I had collected. I laid out a suit, and this was novel: my campaign wear had sunken, over time, to such a casual nadir that Bush and his aides probably wouldn't have batted an eye if I'd started to appear at rallies in Hefty garbage bags cinched (loosely) at the waist. I even made sure that the suit wasn't wrinkled—that it didn't need to be hung in the bathroom and steamed by my morning shower. I was out of control.

But when the former president stepped out of his golf cart, he instantly put me at ease with conversational niceties. He mentioned his recent and imminent travels and his plans to play eighteen holes later in the afternoon with Jeb, who was up north for a visit. A comment of mine about the gorgeous setting elicited a recollection: he said that Sebastian Junger, in his book *The Perfect Storm*, had written that the damage wrought on the New England shoreline by winds and tides had pleased no one, except perhaps a few Democrats because the Bush's house had taken a hit. "So I wrote him a letter," Bush said, "and said, 'Dear Steven,' or whatever his name was, 'I can't believe anybody would be happy . . . I got back the nicest letter from him."

Bush, his personal assistant and I stepped into the house and he pointed to the powder room: "Anybody want to freshen up?" He looked through a window and spotted something that made him smile: a twitching, fascinated springer spaniel. "There's Sadie, our dog, looking for a chipmunk that comes out of that hole and is eating Barbara's gardens alive," he said.

"I think," he continued, "she now has six confirmed kills to her credit." Then, after a second's pause, he added, "Sadie," apparently to make sure I had not thought he was talking about the fearsome Barbara.

He also made clear, although perhaps not intentionally, that not all of the attributes George W. Bush had inherited were ideal. The Texas governor's relationship with the spoken word was a bequest from his father, which became evident after the former president and the former first lady took their positions on separate pieces of furniture in their living room and he began to bemoan the way journalists sometimes wrote about his eldest son.

"The other favorite word that they got is gravitas," George H. W. Bush said, rifling through and trying to rebut the various criticisms that had been directed toward George W. "It's a Latin word."

"Meaning?" Barbara Bush asked, curious to see where he was going with this.

"Meaning, gravitas," her husband responded. "And the answer to that is, I mean, here's a guy that's—talk about an academic record: graduated from two of the greatest—three of the best—schools in the country. And that alone ought to be—I was Phi Beta Kappa graduate of Yale. Nobody said, 'He's got a lot of gravitas. He's a résumé candidate.' It's a fallacious argument. And that stupid little question—was asked all these different, pop quiz here, I couldn't answer those questions and I'm supposed to know something about foreign policy from having been around. I couldn't answer them today and I couldn't answer them when I was president."

"Couldn't pronounce them," Barbara Bush added helpfully.

"No," he agreed. "So suddenly then there's this, well, 'He doesn't have the gravitas.' That's crazy. Absolutely crazy."

Barbara Bush talked less, her interjections coming most often when the subject was how President Clinton had behaved in office and the couple's shared belief that people's respect for the presidency had, as a result of this behavior, waned during his two terms.

At first, she said she was going to resist the temptation to say anything pejorative about Clinton.

"We took a vow," she said. "Can't break it. Never to speak badly about your, your . . ." Her voice stopped here, short of the word "successor." It was as if she was unable to utter it.

"We don't get into that because he beat me and then it looks like a poor sport," her husband rushed in, explaining why they were determined not to do what they nonetheless did, at least to some small extent: rehash what had happened in November 1992, when voters sent him packing after one term and gave a younger, more politically agile man the key to the White House. The Bushes had not left 1992 completely behind them, and this seemed an essential part of understanding why George W. Bush was on his current path.

"You can't blame it all on Clinton," the former president told his wife, referring to what they saw as the diminished station of the presidency.

"Oh, I can," she responded tartly.

"You said you weren't going to," he chided, referring to her vow to speak no evil, "and here you are."

He invoked Clinton to defend his son against charges that the Texas governor knew too little about foreign policy and too little about the world to become president.

"He knows every bit as much about it as Bill Clinton did," the former president said, and then later, unprompted, returned to the subject, this time casting his net wider.

"I'm just going back in my mind," he said. "You had Reagan, Carter, Clinton. What was their foreign-affairs experience? They got elected president."

Barbara Bush then chimed in to imply that Gore's foreign-policy experience was not all positive, alluding to his awkward toast, during a visit to China in 1997, with Prime Minister Li

Peng, who had ordered the violent crackdown on protesters in Tiananmen Square in 1989.

"I don't think we ought to think of Al Gore in China at the time that he was caught with the champagne glasses," she said. "I'm not going to think about that."

"What are you doing?" her husband asked, acting displeased with her caustic interruptions without ever really seeming so. "You're not going to be in this interview if you're going to start talking like that. George will call and he'll be furious."

When we moved from the living room onto the oceanfront terrace out back for a lunch of lobster salad, he sat on one side of me and told her to sit snug on the other side.

"You don't want me to anymore," she complained, "the way I've been talking."

And she kept quiet from then on, except when I specifically turned to her to ask whether—during those wild, aimless years of her eldest son's that were constantly being dissected and researched—she had ever been concerned about him.

"Did he have a picture taken nude dancing?" she said, trying to anticipate my next question and touching on a widely circulated—and never substantiated—rumor that someone possessed such a snapshot. "The answer is, if they did, where is it?"

"And is it a frontal shot?" her husband added, laughing.

It wasn't all like that—wasn't all George and Barbara as George and a version of Gracie with stiletto talons. They showed themselves to be nervous in the extreme about the presidential race, adoring parents in the thrall of a nail-biting wish that their son do well and get what he wanted. This desire, they insisted, was divorced from ideology or politics.

"You wouldn't believe this," the former president said. "I literally don't go into the analysis of these issues. If you asked me, life or death, 'You've got to tell me how big the tax cut is that George is proposing,' I couldn't tell you. I don't give a damn."

Earlier he had asserted: "It's not a question of liberal, conservative, Republican, Democrat. It's a question of pride in one's son. That's what the whole thing is. . . . Our joy is just seeing him and, of course, Jeb, who's up here, and the progress they've made, and now George is in this huge quest and sweating every inch of every mile he goes." As Barbara nodded her agreement, her husband remarked, "People say, 'Well, you know, what's it like?' I say, 'How would you feel if your son is one of two people that was likely to be president?' Forget the fact that I was president and she was first lady. I mean, I wish you saw this place when the grandkids were here and the boys and Dorothy (the couple's only daughter) were here and it's alive. It's all family—everything."

It's all family—everything. This could have been another motto for his son's presidential campaign, and it would have been true in so many ways above and beyond the former president's intended meaning. Passively and actively, indirectly and directly, the Bush family had set this candidacy in motion. It had given George W. Bush both the belief that he could pursue something this big and the need to do it. It had given him the connections, the head start. It had shown him the way, through Poppy's example, which he mimicked throughout many of his years, his life sometimes looking like an old-style carbon copy of his father's: same text, less crisp and commanding rendition, blurrier and fainter and not quite right. The family, its ethos and its experiences dictated the way he ran his political career, especially the loyalty he demanded—and returned—to the advisers in his inner circle. It even, by any reasonable reckoning, had a clear impact on a big, surprising and exceedingly curious decision that would be revealed in late July, just three weeks after I sat with his parents at their oceanfront idyll.

George W. Bush had lived since childhood with the burden of the Bushes' formidable accomplishments, and he had dutifully and worshipfully followed his father's path, first to Andover, then to Yale, then to the oil business. Every step of the way, he never quite managed to match his father's success, and one point of comparison said it all. His father had been a fighter pilot in World War II and had a medal to show for the time he was shot down over the Pacific. More than a quarter century later, George W. signed up to be a fighter pilot in the Texas Air National Guard, avoiding combat in Vietnam and residing, during the final stretch of his service, at a singles apartment complex in Houston, the Chateau Dijon, with a volleyball net strung across the outdoor pool.

He had to know that it wasn't the same, and for all his lightheartedness, there were scattered signs that this awareness bothered him. In 1973, he took his younger brother Marvin out on the town in Washington, D.C., drank copiously and, as he drove the two of them back to their parents' house, plowed into a neighbor's garbage can. His father was waiting for him inside the Bush's house, and when he got there, he did not submit docilely to a lecture or reprimand.

"You wanna go mano a mano right here?" he asked his father, inviting him to step outside. Maybe it was just the alcohol talking. Or maybe it was the nagging frustration of a young man who had come to realize how hard it was to measure up and come to worry about how easy it was to disappoint.

This worry may well have been a part of Bush's decision to quit drinking when he did, in July of 1986. He was turning forty, a milestone that tidily lent itself to new resolutions and significant changes, and some of his and Laura Bush's friends have intimated that she either gave him an ultimatum about his drinking or came close to it. But there was something else interesting about the timing: his father, then the vice president, was revving up for a run at the presidency in 1988, and this meant greater scrutiny of the entire family and a much steeper price for any embarrassments. George W. Bush realized this. He did not want to do anything to hurt his dad, and the kind of blunt, loose talk that his drinking sometimes unleashed was a definite danger. The black sheep, which is what he had called himself, began trying harder than ever to change his hue.

He worked full-time on his father's successful 1988 campaign, then helped only sporadically with his father's unsuccessful 1992 reelection effort and emerged from that loss feeling hurt and angry. "In retrospect, those were such strong emotions for me, to see a good man get whipped," he told me in December of 1999. In fact, they were potent enough that he needed a physical outlet for them, a way to grind out the frustration and dissipate the sadness, so he abruptly and dramatically increased the distance he ran every day, setting his sights on a marathon a mere one month away. Pushing too hard and wearing his body down, he got sick and missed that goal, then bounced back in time to line up for a 26.2-mile run in Houston in January of 1993, just two months after his decision to go the distance—a stunningly short training schedule. It was the first, last and only marathon he ever ran. When I talked to him about it nearly ten years after the fact, he could still recall his parents' reactions as they stood at the 19-mile mark and watched him go by. "There are some elderly women ahead of you!" his mother shouted. "Why are you running so slow?" His father cheered, "That's my boy!" George W. Bush was then forty-six.

Two years later, Bush was being inaugurated the governor of

Texas. In his autobiography, *A Charge to Keep,* which was primarily written by Hughes, he singled out a present he got from his father as a highlight of the day he took office, remarking that the letter that went with the gift still "brings a lump to my throat." He quoted it as saying:

"These cufflinks are my most treasured possession. They were given to me by Mum and Dad on June 9, that day in 1943 when I got my Navy wings at Corpus Christi. I want you to have them now; for, in a sense, though you won your Air Force wings flying those jets, you are again 'getting your wings' as you take the oath of office. . . . You'll do just fine. You'll be a strong, honest, caring governor. . . . Maybe it will help a tiny bit to know that your mother and I will be at your side through good times and bad—right there, always, with total confidence in your ability, and with an overflowing sense of pride in you, our beloved oldest son, and in our wonderful Laura, too. You've given us more than we ever could have deserved."

Reading this, Bush must have felt terrific. And he must have felt better still when he won a second term as Texas governor by a landslide and suddenly found himself one of the most talked-about contenders for the Republican nomination for president in 2000. Despite the profound reservations that Laura had about a presidential campaign, and despite his daughters' fears for the impact on their teenage lives, he plunged in.

Was it some sort of family revenge that pushed him, a desire to take back the White House from the people and the party that had wrested it from his father? Was it an attempt at redemption, a bid for restoration? These were journalists' favorite overarching questions, and any reporter or commentator who answered them definitively was being cocky, gullible, dishonest or some combination of the three, because his or her insights were dependent on sources with axes to grind, interests to promote, truths to conceal, lies to tell. I will give you my sense of things, and it is only a sense. Bush's presidential campaign was not overtly and consciously about any of those "R" words and concepts, which did not ricochet as such around his brain. And the various family members, converging on Kennebunkport, did not look up from the backgammon table and giddily plot the ouster of the renegade Clintonistas as the necessary mechanism to lift the Bushes to renewed glory.

But there were trace elements—or maybe anagrams—of these motivations bubbling beneath the surface of Bush's candidacy, even if he didn't express them or think about them in those terms. When he and his parents said, as they did, that he would restore honor to the presidency, they meant that the Bushes owned that quality and the Clintons did not. They meant that people like the Bushes knew how to respect institutions and traditions, and people like the Clintons never did or would. Of course there was residual bitterness and of course there were lingering recriminations in this perspective. There was also a strain of moral arrogance.

The Bushes were indeed an exemplary family, with values to respect and emulate. Their loyalty to, and support of, one another was unquestionable; any time I listened to one of them talk about the others, I could not fail to hear a love and admiration that transcended the competitive instincts they all had. And any time I found myself in a casual conversation about the Bushes with someone who had known or worked for them, I would inevitably be told a story about the funeral they moved heaven and earth to attend or the sickbed they took pains to visit or the supportive phone call that they didn't really have to make but went ahead and did. Both George the elder and his firstborn son could be touchingly polite and kind in personal interactions, asking after a person's loved ones, putting a person at ease and behaving like the antitheses of the celebrities they were. In their political lives, they could be as pitiless or manipulative as a situation demanded. In their private lives, they seemed truly intent on being as decent and honorable as they could.

Even so, their belief in their intrinsic goodness and fitness for leadership as a kind of premise that needed little explication or elaboration was striking. Writing about the 1992 campaign in her memoir, Barbara Bush says, "I did wonder sometimes in the White House if it was worth awakening every day to the abuse that opponents and the press give the President. I knew that George was the best." In the same book, she publishes one of her diary entries from that year: "I do not know whether George will survive this political year or not, but I do know that he is the only stable person running."

Like mother, like son. In his own autobiography, George W. Bush writes about one of the triggers for his decision to run for president, a sermon, delivered by a minister named Mark Craig, that he listened to on the morning before his inauguration for a second term as Texas governor. "He talked of the need for honesty in government; he warned that leaders who cheat on their wives will cheat their country," Bush remembers. "The minister said that America is starved for honest leaders . . . starved for leaders who have ethical and moral courage." Bush then recalls that his mother, sitting with him during the service, later told him, "He was talking to you."

Bush's most ardent political helpmates adopted this same thinking, and sometimes acted as if Bush-ness itself was proof of civic righteousness, of the impossibility of wrongdoing. In Seattle one night, I sat with a Bush adviser just a few days before the publication of a controversial book—later pulled by its original publishers from store shelves—that made an explosive but wholly unproved accusation about Bush's past. The book's author, J. H. Hatfield, suggested that Bush had been arrested for cocaine possession in the early 1970s and that his work around that time for Project PULL, an inner-city group in Houston, was a substitute for criminal prosecution that his father had negotiated with a local judge. I had seen an advance copy of the book, as had my editors, and we wanted to make sure we talked thoroughly with the Bush campaign about this story, just to be safe. This is what I was doing with the adviser that night, and as he shot down various aspects of the charge, exposing the gaping holes in its plausibility and logic, he concluded with this flourish: "Do you really think President Bush is capable of that?"

Did I think Poppy was capable of saving his son from a felony record by having a little chat with a local judge, by pulling a few strings? Yes, I did (although I wasn't at all persuaded that anything of the sort had happened here), and I did because I thought *most* loving fathers would be. But the adviser played this question—this notion—as a kind of trump card, and the assumption embedded in it was that former President Bush had an unwavering integrity that was above reproach and that I would automatically see, understand and be swayed by this. I found this odd, and I found it extremely presumptuous.

And I could not stop thinking about it when word began to leak out, in late July of 2000, that George W. Bush had selected Dick Cheney as his vice presidential candidate. It's easy to forget now how shocking this was, or how shocking it should have been. Cheney had been assigned months earlier to supervise the process by which a list of possible candidates was constructed, reviewed and later winnowed. For him to emerge as the winner was like one of the female judges of the Miss America contest leaping up from the orchestra section and grabbing the tiara for herself. But it was also shocking because Bush, the self-styled "different kind of Republican" who wanted to broaden the party's base and appeal, had paired himself with someone who was older, whiter, wealthier. Someone who shared his passion for fishing and his ties to the energy industry. Someone who essentially shared his residency in Texas, although there was that other home in Wyoming that allowed Cheney to change his voter registration at the last minute and avoid any constitutional quandaries. Someone who had *already been there*—in Poppy's administration, no less.

Cheney, in truth, was not Bush's first choice, according to several advisers. Number one was Colin Powell, who had no interest in the job and whose associates proceeded to rankle the Bush team by making this clear even before Powell had been asked. (Those associates simultaneously made clear that Powell would prefer, and be prepared to accept, a position as secretary of state.) Number two was retiring Florida senator Connie Mack, who also had no interest, and it's fascinating to wonder how differently the election might have ended if Mack had been on the ticket. He could easily have increased Bush's vote total in Florida above any point of dispute.

With these two out of the race, and with the waning of Cheney's initial resistance to the possibility of being considered, he emerged as the front-runner, to the concern of a few of the Bush aides most personally invested in "the different kind of Republican" dimension of his candidacy. One Bush adviser told me that Mark McKinnon tried to raise questions about the Cheney selection as he and Bush, sometime running partners, put in a few miles at Bush's ranch. Bush's facial expression and lack of responsiveness told McKinnon that the topic wasn't up for discussion. Other aides, too, had their qualms headed off at the pass by the clear signal that Bush was following his own instincts and impulses, thank you very much; strategists looking at the big political picture were not all that welcome to weigh in on this particular brush stroke.

But everyone got on board to try to sell the choice to journalists and voters once it had been made, and it was the manner of this pitch that reminded me of the night in Seattle. In his aides' stated (and, I think, genuine) view, Bush had selected an impeccably "good man," and the proof, which we were all expected to understand and be instantly impressed by, was that Cheney had served as defense secretary under former President Bush. This was the salient fact. This was the irrefutable evidence. He had long been loyal and helpful to the Bushes, so he was clearly decent and honorable. More than this, he had stood beside former President Bush during the Persian Gulf War, which the elder Bush had always viewed as a high point of his presidential leadership and which was suddenly and belatedly being cast as Normandy with camels.

The younger Bush's aides had another favorite method for marketing the Cheney-as-vice-president story. They said that when Cheney had agreed to supervise the selection process, he did not want to be considered for the job, but as he spent more time around Bush, he developed a confidence in him and excitement about his plans that inevitably melted his resistance. To hear this was to wonder whether a significant part of what led Bush to pick Cheney was that Cheney, in saying yes to the vice presidency, was validating him in a way that few if any other running mates could. More to the point, Cheney was validating him in terms of the man who had always been his role model and impossible standard: his father. Cheney's handiest point of reference for the presidency was Poppy, and Cheney was agreeing that the next George Bush was fit for the same post. Cheney was further agreeing to show the younger Bush the same fealty he had shown the elder Bush—in a sense, to invest in him an identical kind of faith, to put him on a similar footing.

The younger Bush would scoff at this analysis as the mangiest kind of psychobabble: he has never believed that motivations are as veiled, profound or complicated as some of us would make them out to be. But there were many signs and clues that Bush still, at some level, might have felt that he had something to prove to his parents, something to prove to himself.

In Kennebunkport in early July, much of his parents' conversation revolved around their surprise at how far George W. Bush, of all the Bushes, had come, a progress they attributed in large part to timing and luck. It was the more studious, more articulate Jeb whom they expected to be a future ruler; George W. was the court jester. Writing about Jeb in his autobiography, George W. concedes, "He was the brother who was supposed to win in November of 1994," when Jeb was running for governor of Florida and George W. was running for the same post in Texas. George W. notes that on Election Night that year, when he won and Jeb lost, "Dad spoke for the entire family . . . when he told the press, 'Our heads are in Texas, but our hearts are in Florida.' "

Now, George W. was the one about to be formally nominated for president by the Republican party, and he was ahead of Gore in polls, and his father offered an explanation that had little to do with his son's intrinsic merits, that dwelled on forces outside his control.

"I think it's change," the former president told me. "I think I was the victim of people wanting change. Eight years of Reagan, four years of Bush, 'We want a change.' And all the things that I felt strongly about never materialized as issues. And I think the same thing is true now. I think people kind of like change, like the idea of change from time to time. I think that works for George's benefit. I don't think Gore can present himself as a candidate of change.

"I'm amazed—still amazed—at the way he's done," his father continued. His mother confirmed that when her oldest son had announced his intention to challenge Governor Ann Richards, an incumbent with an outsized personal popularity, in 1994, she had told him it was an ill-advised, ill-fated idea. The former president and former first lady both talked about how their son had gotten from there to here and conceded that once he had taken the reins in Texas, he was bound to be considered as a presidential possibility. "It's a six-inch putt," said the former president, indulging his fondness for sports—and especially golf—metaphors.

Jeb Bush, who joined us for lunch on the terrace, talked about how much his brother George was benefiting from the Republicans' exile from the White House for two terms, just as Clinton had benefited from Democrats' exile for three terms before that. (As Jeb did this, his mother made a special point of praising how articulate he was; his father had already tossed several of my questions to Jeb, claiming that Jeb was the one who would know best.) Jeb said that the Republican desire to get the White House back was keen enough that it transcended differences between various factions in the party and kept them at relative peace.

"Eight years in the wilderness allows for a higher tolerance for diverse views," he noted as he bit into some lobster, which was permitted on the low-carbohydrate diet he said he was following. "If you looked at Bill Clinton's record as governor of Arkansas, he was given a huge break by groups that were anxious to have a Democrat in the White House. He had a horrible environmental record. . . . He had no distinguishing features in education, yet the teachers' unions were strongly supportive of him because they wanted—they were willing to forgive to have someone, a Democrat, in the White House. Well, I think the same thing applies this year and benefits George, rather than Gore."

The diminution of Clinton was interesting; like Barbara Bush's scattered asides, it suggested that 1992 indeed lived on. But more intriguing was the diminution of George W. Bush, and it warranted notice, because his parents were longtime public figures who knew the power of words and the care to be taken around journalists. They knew to keep their lips somewhat zipped and to steer their comments clear of anything that might yield a headline or a few paragraphs that were anything less than ideal for their son, anything other than laudatory. So if people like them let loose with even a few suggestions that their eldest son had surprised them and done more than they had ever expected—if they let this much sneak out the door—there must be more, much more, where it came from. There must be a powerful belief and feeling of that kind.

Part of his father's take on his eldest son's success may have had to do with the elder Bush's famous discomfort with making things too personal, with anything that carried a hint of vanity. Many of the elder Bush's oddest sentences over time had arisen from his deletion of first-person-singular pronouns, and his version of his son's political ascent hewed to a similar desire to deflect personal credit—his or his son's—and bow instead to destiny. Even so, his remarks suggested a limit to the faith (which is different from love) that George W. Bush's parents and siblings had in him. George W. Bush must have sensed this—must have spent a lifetime sensing this. He must have noticed the kind of thing that I did when, in Kennebunkport, his father deferred to Jeb and then beamed with admiration as Jeb spoke so authoritatively and fluidly. How often, before 1994, before now, had George W. been on the receiving end of such an impressed expression?

Months later, he made a throwaway comment to me that spoke to the same issues. By this point the Florida recount was over, he had been elected president and his inauguration was little more than a week away. He, I and David Sanger, the *Times* reporter with whom I was teamed on the White House beat, were in a parlor in Blair House, which sits across Pennsylvania Avenue from the White House, and is the official guest quarters for incoming presidents and similarly high-ranking dignitaries. We were doing a kind of pre-interview for a longer interview to take place on Bush's ranch in Texas the following day, and it lasted little more than five minutes, revolving mostly around an exchange of pleasantries. As Bush was getting up to leave, he asked me, "How's your old man?"

He meant my father, whom he had heard me mention occasionally on the campaign trail. Once, when he had asked me what I would be doing over an upcoming holiday weekend, I said I'd be traveling to the New York area because it was my dad's birthday. He then whipped out one of his official Governor of Texas cards, scribbled a "Happy Birthday" note, signed it and handed it to me. This was the kind of gesture that mattered to him and that he never forgot, and his inquiry at Blair House about my father was probably triggered by a memory of it. I answered that Dad was doing fine. Bush then remarked that I should tell my father that "the president-elect" had sent his regards.

"That," Bush said, "ought to *enhance your standing* in the family."

There were additional reasons, beyond self-validation, that surely factored into the Cheney choice, but a few of these, too, had to do with family, the constant in—and explanation for—almost all the facets of Bush's life. His father's reelection defeat in 1992, the year that weighed on his and other family members' minds, had for George W. Bush been a parable of insufficient loyalty, in which too many deck hands jumped ship when the going got tough, making it all the more difficult for his father to ride out the storm. So he was determined to have the kinds of sailors who would lash themselves to the mast. Cheney fit the mold—self-effacing, steadfast, unlikely to be contemplating his own presidential run down the line and to be watching out for his own glory. This manner and his history with the Bushes suggested that he would always be loyal, and loyalty was the precious coin of George W. Bush's political realm.

There were also suggestions that Bush's father had been actively involved in promoting Cheney for the vice presidency, in brokering the marriage between him and George W. Bush, but the former president at least officially and publicly denied this. After I wrote an article pointing to evidence of his role, his chief aide, Jean Becker, who had been present in Kennebunkport, called to tell me that I was reaching the wrong conclusion. I suggested that this assertion would carry more force if it was coming from the former president's own lips, and ten minutes later, a number with a Maine area code flashed across the display screen of my cell phone. The elder Bush was calling.

"What I see emerging," he said, referring to news coverage, "is that the old guy drove the choice. That is absolutely inaccurate." But then, responding to questions, he acknowledged that he had offered his son opinions on various vice-presidential possibilities, including Cheney. Asked whether he had spoken directly to Cheney to urge him to think about taking the job, he said, "I cannot reveal whether I did or didn't. Not wanting to obfuscate, but put this down to obfuscation." After a beat, he added, mischievously, "Please respect my age—seventy-six. I'm forgetful."

He digressed to complain about a column Maureen Dowd had just written that joked that Cheney would be baby-sitting the younger Bush, and he reminded me that I had spoken fondly of Dowd in Kennebunkport. The truth was that he too liked her, and respected her, but her pinpricks still drew blood. He said he was thinking about mailing back to her a T-shirt she had given him—it said "Broccoli Lover"—at a recent roast for Marlin Fitzwater, his onetime press secretary. When I relayed the conversation to her, as he clearly intended me to, she told me I should point out that she had referred to him as "classy" in a column. I e-mailed her response to Becker, who passed it along to him. In less than an hour, Becker sent me his written response.

"Dear Jean, OK, she did say something more pleasant than yesterday's horror; but what has she said about me (us) (him) lately that is pleasant?" he mused. On the one hand, he wrote, he wanted to return the T-shirt "unwashed, a terrible punishment indeed." On the other, he wanted to be charitable. "Could it be that at 76 I am looking over the great divide thinking of what 'He'd' want me to do?" he asked. "Would St. Peter say, 'Turn the other cheek and let her take a bite out of that one,' or would he say, 'Return the unwashed tee shirt—an eye for an eye, you know?' OK, I'll think some more. . . . Darn it all. I am now confused!" He never did return the shirt.

As for the question of whether his regular chats with his son centered on formal advice or emotional nurturing, it was ultimately futile to try to make this distinction, because it really didn't exist. Their conversations, at times as often as every day, were about everything and nothing. They were father and son, with a solid if complicated bond that was being strengthened each hour by the fact that the son was going through a life-defining experience that only a finite fraternity of men through American history, including his father, could truly understand.

And his father wanted to be there for him in any and every way, because this was his nature. Back in April, when I had interviewed George P. Bush over brunch in the Brentwood section of Los Angeles, where he was living at the time, he had remembered getting monthly handwritten letters from his grandfather throughout his childhood. One included a sheet of paper on which then-President Reagan had doodled a football player during a meeting; Vice President Bush had got Reagan to sign it. Throughout the campaign, George P. said, his grandfather sent regular e-mails to family members that included nuggets of information from the polls he was carefully watching and repeated bits of wisdom from the politicians and political analysts with whom he had spoken. He was a veritable junkie for the race.

He tried hard to create a public impression of remoteness from it, especially after his comment in New Hampshire about his "boy" and subsequent commentary that he had done his son damage, made him seem less than his own man. Although he had attended the Kentucky Derby, sitting in a sky box where his son joined him for several hours, he did not walk with his son through any public areas, and reporters, film crews and photographers weren't allowed to get a glimpse of the two men together. He encouraged his sons not to be too considerate of his own political past. In a letter that he wrote to both George W. and Jeb in August 1998, and that was printed in George W.'s autobiography, he said, "At some point both of you may want to say, 'Well, I don't agree with my Dad on that point' or 'Frankly I think Dad was wrong on that.' Do it. Chart your own course, not just on the issues but on defining yourself."

His own course? It would be hard to apply that term to what George W. Bush was doing. His course in many ways began and ended in Kennebunkport, which was not just a physical place but a web of historical and emotional realities that tied the Bushes together. He was trying to clear the highest bar that his father ever had, a higher bar than Jeb probably ever would, and he was up there, with his chin just over it, because his family members had taught him that he had a right and claim to such a place, even as they perhaps failed to communicate any expectation that he, among all the Bushes, would reach it. He was up there because it was a view he had never seen before, and it was nice in large part for this reason.

It's all family—everything. Even after George W. Bush succeeded in winning the presidency, the phrase reverberated; so much of what he did seemed to be linked in some way or another to what his father had and hadn't done, although not in the carbon-copy manner of his earlier life. His father, in failing to secure a second term in office, had been accused of belittling the seriousness of an economic downturn for which he was assigned responsibility. The son spent so much time in the first months of his presidency pointing to troubling economic indicators that he was accused of actually talking the economy down, and he was careful in his public remarks to stress his understanding of the exact ways in which Americans were fearful or hurting. His father had infuriated conservatives—and politically wounded himself—by agreeing to a tax increase. The son went to work instantly on getting Congress to pass a substantial, across-the-board tax cut, making the matter his highest priority.

The father had waged and won a war in the Middle East only to confront, years later, a debate about whether he had gone far enough, whether he was wrong to leave Iraqi president Saddam Hussein in power. In the days immediately following September 11, 2001, one of the questions was if Hussein's fingerprints were at all visible and if Iraq had done anything to harbor or educate these terrorists or others like them. And suddenly the son was again revisiting some of the turf that once belonged to his father, trying to incorporate the right examples and lessons and learn from any mistakes. (He even had many of the same lieutenants, principally Cheney and Colin Powell, who had been chairman of the Joint Chiefs of Staff during the Persian Gulf War and was now secretary of state, around him.) The second President Bush set his sights on more than just the band of terrorists responsible for the World Trade Center and the Pentagon but also on terrorism itself and on the countries that fostered and facilitated it. He was not going to be accused of going only half or three-quarters of the distance.

CHAPTER SEVEN

IN LATE AUGUST OF 2000, WITH THE ELECTION two and a half months away, there erupted an argument between the Bush and Gore campaigns that struck to the very souls of the candidates, underscored their disparate visions for the country's future and was fought with a gladiatorial ferocity by their aides. I am not talking about taxes. I am not talking about Social Security. I am talking about something infinitely bigger and infinitely more profound: the question of who covered more miles and more states with more velocity on his initial campaign swing after the political convention at which he was formally nominated for the presidency.

This really happened, and it happened even sillier than I just made it sound. It was Saturday, August 19, the second day after the end of the Democratic convention, and Bush and his aides had woken up to news articles that noted just how unimpressive his schedule on Friday had been. Gore and the Democrats had hogged the spotlight from Monday through Thursday, the duration of their Los Angeles jubilee, just as the Republicans had done weeks earlier in Philadelphia. And on Bush's first day back in the news, he had traveled all the way from Austin to Memphis, Tennessee, for a grand total of—drum roll, please—*one* event. He then beat a hasty path back to Texas to go for an afternoon run on a favorite track in Dallas. It was an itinerary seemingly oblivious to the kind of post-convention bounce Gore was getting in the polls, and it suggested that Bush was either not sufficiently concerned or in a kind of retreat. Reporters took note of it.

When Bush's aides saw the Saturday papers, they were displeased, and so Scott McClellan, one of the campaign's spokespeople, recited for those of us in the press corps a series of important and putatively impressive statistics. According to McClellan, Bush had racked up, on his post-convention train trips through the Midwest and up the spine of California, a total of twenty-two rallies or other events in five states that represented 138 electoral votes. By way of helpful contrast, McClellan pointed out that Gore, on *his* post-convention riverboat trip down the Mississippi, would visit only four states, representing 51 electoral votes, and manage just eight rallies or similar events. One reason for this, McClellan pointed out, was that the trains "hit a top speed of 75 miles per hour compared with seven miles per hour" for the riverboat.

There were, however, a few kinks in McClellan's arithmetic, as calls to the Gore campaign and a closer inspection of McClellan's figures soon revealed. McClellan was providing us with a seven-day estimate for Bush's travels, covering the three straight days that he traveled through the Midwest on a train and the two later days (which followed two days off) that he rode the rails from southern to northern California. McClellan was wrapping it all into one package because the mode of transportation had been consistent, even though the candidate had taken a breather in the middle and changed trains. And McClellan was using, for Gore, a period of only four days, because that was the entirety of the time he would spend on his post-convention vehicle of choice: the old-fashioned Mississippi paddle boat.

Gore's aides were duly horrified—and duly irate. They said that if Gore's movements were measured over a similar seven-day period, his number of events rose to twenty and his number of states to six: one more than Bush's. "Al Gore has been literally campaigning twenty-four hours a day, seven days a week, for a year," his ever-quotable, saber-toothed spokesman, Chris Lehane, said. "We do not take time out of our days so we can . . . sleep with our own pillows, as Governor Bush seems to do." Another Gore aide disputed the count of eight events over the first four days, saying the campaign's schedule showed fourteen.

This was not the end of it. There was a debate between the two sides over what constituted an event: If, say, Gore held a discussion with a small group of voters on the deck of the boat as he plied the waters between one rally and the next, shouldn't this be a part of the tally? Were thumping campaign music, banners, a stationary lectern and a cheering crowd required? McClellan said that Bush's total would in fact be higher if it were amended to include "slow rolls," or moments when the train dramatically cut its speed so Bush could wave to a throng of people who had gathered by the tracks. Wait, hold on a second: Gore's aides let it be known that the riverboat trip had also featured, well, "slow floats," and counting those would similarly increase *Gore's* total. (Of course, if the boat was slowing down from seven miles per hour, it must have been going backward, or sinking, but we never got into that.) Too bad Forbes wasn't still around, running as an independent: he could have rented a zeppelin or hot-air balloon and really made the competition exciting.

The period of the campaign between the conventions and November 7 should have been the most intensely focused on the stuff that mattered: the actual differences in the candidates' positions, the best indicators of whether they would be able to follow through on what they proposed to do. And yet it was, instead, the kookiest, the most rife with diversions, the most redolent with clear signs that the candidates, their advisers and the journalists covering them had lost much of their touch with reality.

Those of us in the news media plumbed new depths of disingenuous behavior, zooming toward anything sexier than issues and then commenting sadly about the way in which issues were being eclipsed, pumping up the importance of something trivial and then marveling at how prominent it had become. The candidates and their advisers scaled new heights of manipulation, their carefully planted phrases and gestures sprouting and spreading like weeds. At the Democratic convention, Gore sought to show that he possessed a pulse, a passion—and, one hoped, a breath mint—by gathering Tipper into his arms and performing surgery on her tonsils with his tongue. Bush couldn't let a kiss like that go unanswered, and so he came back with his own, placed on the demographically felicitous cheek of Oprah Winfrey, the most famous black woman in America and television's empress of empathy, during an appearance on her talk show on September 19. Bush's advisers literally came to see and talk about this as a critical moment in his seduction of voters, a crucial part of his subsequent victory.

I don't know about that. But I do know that Bush's appearance on *Oprah* was a signal of something important about the candidate himself—of his bit-by-bit willingness to acquiesce, to swallow his medicine, to behave. *Oprah* wasn't the kind of thing he was given to doing. It wasn't his bag at all. The show was about heart-baring and soul-baring and the kind of psyche-probing that Bush detested, but several of the women around him—Laura Bush, Karen Hughes, Mindy Tucker—urged it on him, telling him it would be a good thing. And so he relented, and went, and managed to seem, for all the world, as if he enjoyed being there, a feat he had not accomplished during those first, dreadful debates of the Republican primaries, which he had also resisted. The long and grueling campaign was indeed seasoning Bush. It was making him a more obliging and flexible sport about certain things, and it was making him eat his peas.

He still preferred sno-cones, at least metaphorically speaking. On those trains that carried him beyond his black-is-beautiful, hide-Tom-DeLay-in-the-attic convention, he stood on the back platform, the wind whooshing around him, and wore the tickled expression of a six-year-old who had just made the height requirement at Disney World and was riding Space Mountain for the first time. He swung from the vertical poles on the platform, one of which Cheney, standing rigidly beside him, could easily have been mistaken for. He waved joyously at people who lined the tracks, unbowed even, in the sprawling metropolis of Odell, Illinois, by a woman whose pants were pulled down so that Bush could read a message across her buttocks that said, "Raise Min. Wage."

"Congratulations!" Bush shouted to her. "You just got on national TV!"

"You're goddamned right," the woman shot back.

And Bush, in a slightly more chiding and prim voice this time around, reminded her: "You're on national TV." At some point along the way, he had become the grown-up.

He did not flinch, growl or even smirk when, during the California train trip, McCain joined him for a rare dual venture and, in little barbed asides, made clear yet again that he wasn't all that keen on his party's presidential nominee. He and Bush first crossed paths at an event in Salinas, California, in the Central Valley, and when McCain introduced Bush, his three minutes of remarks were perfunctory in the extreme. McCain took advantage of Laura's presence to tell the audience that she was "the brains of the outfit." Somehow, it didn't come across as a stock, throwaway phrase.

Bush and McCain sat together on the train, at one point giving a short joint interview to CNN in which McCain addressed the subject of Bush's preparation for the presidency with a nod to where he and Bush were traveling. "We had another president that came out of two terms of being a governor," McCain said, referring to Reagan and California. "And he did pretty well." Then McCain added, jokingly but maybe not, "You know, I hate these governors. We should only take senators."

Bush seemed positively elated to get off the train and away from McCain when it stopped in Guadalupe, California, a dusty, grimy, impoverished and almost entirely Hispanic town halfway between nowhere and desolation. For a Republican, this was called great television, and the chatter that greeted Bush as he strode down the dilapidated main street sounded like it was coming from hired extras, which I wouldn't rule out.

"Bush is the only Republican who would do this!" shouted one man. "Bush is the only Republican who would do this! Bush is the only Republican who would do this!" I took an extra close look at him, suspicious for a few seconds that he might really be Hughes, Rove, or McKinnon wearing a fake mustache and bronzer.

Bush signed autographs and tapped his reservoir of Spanish phrases to trade pleasantries with local store owners. He dipped into a restaurant called La Simpatia, a movement that had clearly been choreographed in advance, because Laura was seated at the lunch counter, waiting for him. Slowly working the restaurant, Bush made his way to the seat beside Laura precisely as one of the restaurant's owners was placing a heap of nachos in front of her. Bush reached hungrily for the grub.

Then he turned around. "Anybody want a nacho?" he asked several dozen adoring minority residents of the town, for whom this was clearly not about politics but sheer celebrity.

"Anybody want a president?" someone shouted back, and Bush chuckled heartily.

Bush plopped another nacho in his mouth, then emitted a slight choking sound as he struggled to swallow it.

"Don't do what your father did!" someone chirped, referring to former President Bush's unfortunate episode of presidential regurgitation at the home of the Japanese prime minister in 1992. The candidate did not chuckle so heartily this time.

This whole sequence of events lasted at least twenty minutes—longer than most of the remarks Bush was making at his whistle stops that day—and it was all the campaigning he did over about a three-hour span of time as the train chugged further and further north. Anyone who wonders why the journalists covering campaigns get so cynical—and inject so much of this cynicism into their copy and broadcast reports—need only know that this kind of "event" was not at all rare and, on some weeks, more common than anything else.

But our understandable cynicism did not excuse other journalistic tendencies that flourished during this stretch of the campaign. In the hyper-competitive atmosphere engendered by so many reporters covering the same story, there was a near-constant rush to be the first to notice when one candidate was slipping and the other rising, one reeling and the other gathering force. Everyone was looking for "moments," for turning points, for inklings of a changed dynamic, and no one was sensitive enough to the danger of willing such changes into being rather than accurately noting their occurrence. Everyone wanted to come across as smarter than the next guy or girl. Everyone wanted credit, later on, for being prescient.

And so Bush, in late August, suddenly found himself fighting the rap that his candidacy was on the ropes. He had come roaring out of his own convention with a nearly twenty-point advantage in polls, an expansion of a lead he had long maintained. But then Gore emerged from his convention with a slight edge, and it did not disappear, at least not right away. Was this it, then: the beginning of the end for Bush, the widely predicted realization by many voters that Gore was the incumbent, these were good times and there was no sense in turning elsewhere? Many journalists were itching to declare this—not out of sympathy for Gore, but out of a desire for a new twist to the campaign narrative—and toward the end of August, they more or less began to.

On a Monday night, at a fund-raiser in Des Moines, Iowa, Bush gave a speech that turned into a blooper bonanza. He twice warned about terrorists holding America "hostile," when of course he meant "hostage." He intended to say "tariffs" and instead uttered "terrors." He talked about putting a final nail "*on* the coffin" and he suggested that Clinton had been in office for four years, not eight. None of this was unprecedented—these downward verbal spirals came and went. But the next day, Bush encountered two additional misfortunes, one of his own making, that several reporters decided to lump together with his awkward speech. He said on his campaign plane that the softness of public support for his suggested tax cut meant that he had not explained it well enough, then sought to explain it with numbers so divorced from any context that they only begged *additional* explanation. And at a rally that day in Peoria, Illinois, one of the local Republicans, U.S. Representative Ray LaHood, told a gaggle of reporters that many people in Congress thought Bush's tax-cut proposal was too big and probably wouldn't fly.

This cluster of developments, because it came at the same time as the tightening of the polls, produced a flurry of stories that deemed Bush to be in serious trouble and filtered all evidence through this predetermined conclusion. Never mind that LaHood might have said what he did four weeks earlier or later and that we had simply run into him when we did. Never mind that Bush's malapropisms, although worse and more numerous during times of stress, also popped up during stretches of the presidential race when everything was going his way. Now they were considered obvious signs of extreme nervousness, just as LaHood's statements were proof positive that the tax cut was a set of weights on Bush's ankles that would not let him cross the finish line first. Reporters were all but hammering the nails *on* Bush's coffin.

Mike Allen of the *Washington Post* led the charge, turning the aforementioned developments into a front-page story that began: "After five months in firm command of the presidential race, George W. Bush suddenly finds himself on the defensive, behind in some of the polls and struggling to fend off attacks." All of this meant that Bush was facing "his biggest test" yet "as he struggles to adapt to rockier terrain." Allen had spent almost no previous time with Bush—he had initially covered Bradley's campaign—but this lack of context did not stop him from finding and pronouncing Bush's demeanor and confidence off-kilter, and it did not stop other journalists from following suit. There was a "moment" to be had or—heaven forbid—*missed,* and there were tough-minded, keen-eyed reputations to be made.

As for Bush being on the "defensive," a term and observation that traveled well beyond Allen's copy, it was we in the press corps as much as anybody else who had put Bush there. Here is how it had worked: Gore was hurling newly pointed critiques of Bush's suggested tax cut around the same time that Bush made himself available for questions at an impromptu news conference, and we used this opportunity to relay Gore's words to Bush and ask him to defend his position. Over in Goreland, the vice president was not making himself available to reporters, who could not get *his* response to whatever charges Bush was making about him. Ergo, Gore was not on the "defensive," while Bush was.

I deliberately soft-pedaled Bush's supposedly bad two days at first, ignoring the Monday Night Bloopers—I had written plenty in the past about Bush's adventures with English—and dwelling principally on Bush's quasi-admission on Tuesday about the difficulty of selling his tax cut, which seemed, defensive or not, to be an interesting, relevant remark. But within a few days, I too was toughening my language and partly rejoining the pack, because the assertion that Bush was flailing was so rampant in newspapers and newscasts that it had transmogrified into the *fact* that Bush was flailing. It was the given, the starting point from which every story now proceeded. It was true because almost every journalist, following a kind of groupthink endemic to campaign coverage, had decided that it was true and was reporting that it was true; because this was all Americans were reading, hearing and likely to believe; and because the Bush campaign would therefore have to accept and react to it as true, making it all the truer. I wasn't sure if this logic was pretzel or circular, but I knew it was pathetic.

So much of campaign coverage—and of political journalism in general—carried this strain of dishonesty, by which news organizations, having focused on something in particular, then published or aired stories about the problems a candidate was having because of where the focus on his or her activities had fallen. So much of campaign coverage took reporters' assessments of events and presented them as objective, generalized realities. This was certainly the case with the debate about the general-election presidential debates, which, over the course of several weeks in early September, probably claimed more ink and filled more broadcast time than any other aspect of the presidential race.

Around Labor Day, Bush announced that he did not want to do the three ninety-minute events proposed by the nonpartisan Commission on Presidential Debates, which had administered such forums for the last several presidential cycles. He wanted to do just one of them and then two others that would be sixty-minute, television news program–style encounters chaperoned by NBC's Tim Russert or CNN's Larry King. (Bush felt particularly at ease with both of these men, and his aides knew that Gore was not an enormous fan of Russert's.) It looked to those of us in the news media as if Bush, frightened of too many prolonged face-offs with Gore, was trying to reduce his exposure and possibly the size of the television audience and make sure each setting was uniquely suited to George W. Bush's tastes and whims. We wrote this. And we were right to write it.

But we didn't stop there. We found ways to write about it a second and third time, and we turned the stand-off between him and Gore, who wanted the three commission debates, into the main line of questioning we would direct at Bush whenever we had a chance to ask him something. Then—and this is the beauty part—we wrote that Bush was being dogged and hurt by the impression that he was trying to avoid the debates. Our evidence that this impression had taken hold was to be found *not* in polls or extensive interviews with voters—who, for all we knew, didn't care when or how the candidates debated—but in the profusion of news coverage devoted to this impression, which was ours. As for Bush being dogged and hurt, he indeed was, but the pain-inflicting canines in question were reporters and editorialists and pundits. We were, in essence, covering ourselves.

Then along came a rat, and we did it all over again. One morning in mid-September, the *Times* ran a front-page article by Richard L. Berke that noted that in one of the Republican National Committee's television commercials, the letters R-A-T-S, a portion of the word "bureaucrats," was showcased on the screen long enough, perhaps, to send a subliminal message about Gore's health-care plan, the topic of the spot. Republicans said that this was entirely unintended, the by-product of a computer-driven editing technique, and there was some evidence and analysis to support them, although there were also reasons to be skeptical. It was, in the end, a hazy matter—curious, worth noting, not worth dwelling on for all that long. But for the twenty-four hours after the story appeared, Bush was asked about it incessantly, to the exclusion of almost any other topic. And this was the coverage that resulted:

"Bush is knocked off his message, and the timing couldn't be worse," proclaimed NBC correspondent David Gregory, turning to a superlative with apocalyptic undertones. Dean Reynolds of ABC went gentler as he explained the meaning of it all to the anchor Peter Jennings, saying, "For one very important day, the Bush campaign's message was diluted and, at times, overshadowed. It's only one day, Peter, but there are just fifty-six of them left in this race." Sounded like plenty to me.

George E. Condon Jr. and Toby Eckert of the Copley News Service wrote that Bush's "drive to retool his presidential campaign hit a detour," while G. Robert Hillman of the *Dallas Morn-*

ing News wrote that Bush was pushed "off message again as he struggled to retool a campaign." The *Hartford Courant* editorialized that "the 'rats' controversy has overwhelmed the Republicans' message." And you can color me conformist, because I began my story on Bush's activities on the Day of the Rats by writing that "another of his campaign days skidded onto unexpected ground" and that an accumulation of distracting surprises could "threaten to blunt his efforts to regain some of the ground he seems to have lost to Mr. Gore." That was surely the case, but only because reporters, including me, were making it so. Bush would not have been spending a day or two fending off questions about an RNC advertisement—or, before that, about other sideline developments—if the questions weren't being asked. And we were the ones asking them.

With the next prime distraction, we distinguished ourselves anew. Tom Downey, a former Democratic congressman who was playing the part of Bush in Gore's debate practice sessions, retrieved his mail one day and discovered a package, sent anonymously from Austin, that included a videotape of one of Bush's debate rehearsals. He immediately recognized it as a possible example of thievery and turned it over to law enforcement officials, after which point a favorite theory emerged in the Bush press corps—and, astonishingly, in print. It was this: that Rove had in fact sent the tape as part of a "dirty trick" meant to entrap Gore campaign officials. See if you can follow this reasoning, because I sure couldn't: Rove expected Downey to hold on to the tape and actually use it. Then, at some point, Rove was going to alert other campaign officials and the news media that one of Bush's debate preparation tapes was missing and had possibly been stolen from the Bush headquarters. Maybe he would do it after the first debate, citing aspects of Gore's performance that were clearly developed from the contents of the missing tape. Suspicion would focus on the Gore campaign and make the vice president look bad. Bush would then benefit by comparison.

The *Houston Chronicle*, the *Dallas Morning News* and the *Washington Post,* among other publications, all reported and repeated charges by Democrats in Texas that the incident had Rove's fingerprints on it, because Rove was known in Texas as a ruthless, conniving type and because he had once, during a previous campaign, announced the discovery of a surveillance "bug" in his office that he was widely thought to have planted there himself. But none of the stories in these newspapers answered a slew of pressing questions about Rove's intended strategy. Why would people necessarily believe his claim down the line that the tape was missing? Why would he need to send a tape to the Gore campaign to make such a claim, since they weren't going to be stupid enough—once he made his public charge—to keep the tape on an office coffee table for the FBI to discover? And if Rove was so crafty, wouldn't he foresee the possibility of Downey doing exactly what he ultimately did, which made the Gore campaign look ethical and upright?

As it turned out, the FBI investigation into the matter quickly centered on a low-level employee at Maverick Media named Yvette Lozano, who had previously worked extensively and fervently for Democrats. A camera at the post office from which the debate tape had been sent had photographed Lozano on the precise day when the tape was postmarked, but Lozano claimed that she was simply returning, by mail, a pair of Gap pants that her boss, McKinnon, had ordered on-line. It was because of this detail that I actually found myself on the phone with McKinnon one afternoon verifying both his waist size—29—and the fact that he had bought the pants in question on the Internet because his compulsive distance-running had winnowed his frame to the point where the only adult pants in most Gap stores that fit him were women's. This was political journalism at its edgiest.

Lozano was eventually indicted, the following year, for mail fraud and perjury, then pleaded guilty to a reduced set of charges and was given a short prison sentence. I learned from people close to the situation that there were several possible explanations—beyond her affiliation as a Democrat—for why she would betray McKinnon, a longtime friend whose kids she baby-sat. Lozano had apparently wanted to go to the Republican convention with other Bush campaign personnel, but wasn't needed there and was told that the campaign would not pay her way. She stayed home. She had also been using the Maverick Media credit card for personal expenditures, and when this was discovered, shortly before the debate tape was sent, Maverick started docking her paycheck to make up for the money she had spent.

But senior campaign officials didn't bother to look into or learn any of this before they began to insinuate that the FBI, an arm of the Clinton administration, was neglecting the obvious likelihood that someone who supported the Gore campaign had, like the Watergate burglars, sneaked into the Maverick offices and stolen confidential material. And they cast Lozano as the hapless, meek victim of a pernicious persecution. Hughes made a point of referring to her as a "young Hispanic woman who doesn't have enough money to afford a lawyer," even though the reporters Hughes was talking to had not asked anything about Lozano's ethnicity or financial portfolio. (For the Bush campaign, there was never a bad time to work the diversity angle.) Hughes also complained that Stuart Stevens, an adviser who worked with McKinnon on the campaign's commercials, had not been interviewed by the FBI, even though he "actually saw some Democratic consultants late at night in an adjacent production facility."

As Election Day drew closer and advisers' spinning became more fanciful, Hughes was developing a quicker draw, sometimes with embarrassing results. Shortly after Bush finished taping the *Oprah* show, for example, Hughes loped into the room where reporters were filing their reports to let them know what a success the campaign considered Bush's appearance to be. "We are getting incredibly positive feedback," she exulted. "We've talked with people who saw it in different cities all over the United States, and the feedback is overwhelmingly positive. 'Home run' is the phrase most frequently being used." This phrase, in fact, was the clue that she might be improvising just a tad; most Americans, unlike Bush, do not speak spontaneously in baseball metaphors. She left the room, and a few minutes later, Gordon Johndroe materialized for the purposes of making the tiniest amendment to Hughes's remarks. He clarified that her reference to "cities all over the United States" was not exactly 100 percent correct, because the *Oprah* show, taped in Chicago and broadcast at different times in different places, had not yet aired anywhere but its home area. But those Chicagoans, Johndroe assured reporters, had indeed been an enthusiastic bunch.

Bush's statements on *Oprah* were, as the format dictated, mostly tame and trite, and he submitted to the same oral questionnaire that she had given Gore when he had appeared on her show. He named his favorite sandwich (peanut butter and jelly on white bread), favorite gift (a kiss from his wife), favorite fast food (taco), thing he could not live without (running) and prized possession (his Texas ranch). But when Winfrey asked him his favorite dream, he seemed briefly taken aback. He blushed, and he held his hand up.

To some observers, it looked like a gesture that said, "Stop, no, that's getting too personal." And such a gesture might have been consistent with much of Bush's behavior during the campaign, with his refusal to surrender too great a portion of his private time and private self to his ambition. But what he was really doing was miming the act of taking the oath of office, of being sworn in as president. I'm not sure this was, really, a vision that danced ecstatically through his head at night. But he had committed to it, and was chasing after it, and his growing willingness to cast his presidential ambition as bigger than he had when he started his campaign, or even when he was midway through it, was an example of his evolution. Bit by bit, he was trying to get rid of those instances of insouciance that went too far and to leaven the ones that remained with instances of a poignant earnestness. Bit by bit, he was embracing his mission, buckling down to it. As it did in South Carolina, it was happening now at a moment of potential crisis. And it suggested that one part of Bush could overcome the others if he decided to let it—that there was more to him than met the eye.

Advisers considered the *Oprah* appearance so important for different reasons. It came after the debate about debates and the hullabaloo about R-A-T-S and the other distractions, allowing Bush to talk directly to millions of voters without a conduit of journalists—a "filter," he liked to call it—that decided to jettison 97 percent of what he was saying. *Oprah*, with its huge female audience, also represented a more targeted attempt to close the "gender gap," by which women disproportionately favored Democrats, and Bush soon pressed his wife and mother more fully into that effort. Laura Bush and Barbara Bush teamed with Condi Rice and Lynne Cheney and fanned out across Midwestern swing states for several days on a "W. Stands for Women" bus tour. Desperate for a change of oratorical scenery, I went along for the first twelve hours, after getting a guarantee that I could spend thirty minutes alone with the two Mrs. Bushes.

I had seen and talked with both of them before, and I found Laura particularly intriguing, because she remained, to an outsider, so completely and steadfastly unknowable. I had never met someone in public life who seemed to have so little appetite for it and brought so little ego to it, and I had seldom met someone in public life who was so consistently wary around reporters. She fell back time and again on the same words to describe her feelings about things, to the point where the words held no meaning at all. Anything that happened to her—being the wife of the Republican nominee, giving a speech at the Republican convention on live national television—was simply "fun," or "really fun," or "really a lot of fun." Later on, when she and her husband were finally ensconced in the White House, an interview she gave to the *Washington Post* perfectly captured this tic. She was quoted on her experience in the job of first lady ("really fun") and her initial impressions of George W. Bush ("very fun," "very funny in a fun way," and "a lot of fun"). About five months into the first year of the administration, I ran into her at a party and, knowing how much she had dreaded the isolation of life in the White House, asked her how she was making the adjustment. She said she was having a lot of fun.

The very first time I had a real chance to mingle with her, outside the governor's mansion in Austin during a Halloween event in October of 1999, I mostly embarrassed myself. There were only a few of us from the press chatting with the first couple of Texas, and Peggy Fikac of the *San Antonio Express-News* and I took seats near Laura on a lip of asphalt beside the front walk. Fikac had previously worked for the Associated Press, and before she left that job, she reported and wrote a profile of Laura that died along with her resignation. Laura quizzed Fikac about why it had never run, but I missed that part of their interchange and tuned in only to hear Fikac talk in general about situations in which profiles were killed, remarks that I did not guess had anything to do with Laura. I decided to throw in my two cents and try to enliven the conversation.

"Well," I said cheerfully, "I can tell you why a Sunday magazine story that I reported and wrote about a year ago never got published." I then proceeded to recall a certain politician who had insisted that almost all of our interviews be off the record and was so careful and uninteresting on the record that there was simply not enough good material to justify the ink. Hearing this explanation, Fikac turned pale. Laura's smile dimmed a few watts. And since I had not really listened to or absorbed the conversation that had preceded my helpful journalistic tutorial, I had no notion why. Fikac later explained the situation to me, at which point I weighed the merits of moving to a new country versus simply changing careers.

I chose a third course: to be as cordial and engaging around Laura as I possibly could. The problem with this approach, in addition to the limited number of weapons in my arsenal, was that Laura did not appear to be at all interested in being charmed by anybody in the press. I remember in particular a small party for supporters that the Bushes held at a hotel in Scottsdale, Arizona, following a debate in Phoenix in early December of 1999. The Bushes had reached the party before reporters could catch up with them, and by the time several colleagues and I entered the room, it looked as if the Texas governor had already made his remarks, because he was weaving through the crowd, shaking hands. Laura was standing several yards from him, more or less alone.

I approached her. "Hello, Mrs. Bush," I said, my voice as honeyed as a hive.

"Hello, Frank," she responded.

One beat.

"Well," I said, "did the governor already speak?"

Two beats.

"He did."

One beat.

"Well, we must have just missed it."

A beat and a half.

"You did."

An infinity.

"Well," I said, leaning anew on this one-syllable space filler, "did he make any news?" This was a joke—and, I think, a fairly obvious one. A candidate never tries to make news on the night of a debate, because the debate is the only story journalists are going to cover.

Laura answered it straight. "He said that he feels good about his chances of winning Arizona," she said, referring to the late February Republican primary there. It was probably a faithful rendition of his remarks: he said the same thing about any state he visited.

"You're kidding!" I responded in what I believe, to this day, was a transparently facetious, merry, let's-have-a-laugh-shall-we vein. "I thought he had already *conceded* Arizona!"

"No, no," Laura corrected me, dead seriously. "He's excited about his chances here."

At this point I made for the buffet table, which had a spread of corn chips and *queso* that suddenly looked much more appetizing than it had just minutes earlier.

But even on this awkward night, my antennae told me Laura was neither dense nor humorless—just guarded in the extreme. She also, it turned out, was nowhere near as square as her unimaginative, conventional outfits and unimaginative, conventional haircut suggested. When she and a group of friends staged joint celebrations for their fortieth and fiftieth birthdays, their chosen recreation was river rafting. Republicans in Washington and arts-and-music types in Austin said that she had quite a number of gay male friends. Young people who worked in or around the campaign said that she would sometimes take an interest in their romantic lives that definitely fell on the cool side of motherly. Others who crossed paths with her during the campaign said that sometimes, when she was in social settings or other surroundings where she felt totally safe, she would bum cigarette after cigarette, indulging a minor vice she never acknowledged in public. (In fact, during an interview on CNN after she became first lady, she flatly denied it.)

She had a bit of a sassy streak that came through now and then and seemed to show itself in a comment she made about her mother-in-law in one of my first interviews with her. Laura was talking about that 1993 marathon that her husband had run, and she was remembering that he and she had stayed, the night before the race, in the Houston home of his parents. In the tradition of carbo-loading, Barbara Bush had made a dinner of pasta, Laura recalled. Then Laura added, with the slightest of edges in her voice, that it was probably the first time Barbara had cooked in more than a decade.

Laura's frequently noted interest in books—she had once worked as a school librarian—actually bordered on a serious love of literature. Long after she left graduate school, she took recreational classes on the novels of William Faulkner and on Greek mythology. Every few years, she reread the "Grand Inquisitor" section of *The Brothers Karamazov.* During the Republican primaries, she was on a Truman Capote binge, but found herself in the middle of *In Cold Blood* as her husband was turning the corner from his loss in New Hampshire to his hopes for a recovery in South Carolina. With all this real-life anxiety enveloping her, she found she had to put the book down.

She had sadness in her past, and it was tempting to conclude that certain habits in her present were a reflection and response to that. When she was growing up in Midland, Texas, her parents tried repeatedly to have another child, and several times her mother either miscarried or gave birth to a baby that died in infancy. Laura told me she was old enough to know what was going on, feel their pain and try not to add to it by ever failing to please them. "I didn't want to upset them in any way," she told me. Then, when she was seventeen, she drove through a stop sign and barreled straight into the car of a classmate from school. He died. Now, as a grown-up, she was a cleanness-and-neatness freak, scrubbing and organizing all the time, inoculating her world against disorder, eager to keep everything pristine and in its place and all the lines and angles just right. The books in one of the houses that she and George had owned had been arranged in accordance with the Dewey decimal system. She liked her skirts to fall *just so* on her leg and whenever she tried on a new outfit from the boutique in Dallas that designed clothes for her, she knew instantly if the length had been tweaked by even an inch in one direction. And she disapproved, albeit graciously.

She was mostly but not entirely traditional, a woman who chose what she knew was a gender-typical career by going into education but also remained unmarried—and on her own—longer than any of her friends. They said that while Laura was pretty and sure of herself and never lacked for dates, she felt no compulsion to be part of a couple. And yet, after she married George W. Bush in 1977, at the age of thirty-one, she quickly gave up a career, dedicating herself to him and their children. She had little demonstrable yearning for anything else, and said over time that simply putting the children to bed or puttering in the garden were the precious elements of a blissful life.

Her mother-in-law was easy to understand, and consistent in her benignly malicious (or were they maliciously benign?) ways.

Barbara Bush knew, for example, to smile and laugh innocently while eviscerating someone. Talking with me about Laura during a telephone interview, she said that if Laura became first lady, she would be nothing like Hillary Clinton.

"I think she would rather make a positive impact on the country," Barbara said. "And I'm not criticizing Mrs. Clinton. But it's like oil and water. We're talking about two different subjects. They're two different people. I think Laura thinks of others." Then, to make this observation seem less venomous than it had been, she tore into herself by predicting that Laura would also be "better, much better" in the role than she herself had been, because Laura wouldn't "be a fat, white-haired old lady." Barbara Bush was one shrewd operator.

Laura was more of a mystery. I interviewed her in the governor's mansion one day and she was talkative and, well, fun. I interviewed her on the campaign plane a week or two later and she was reticent and lifeless, and as soon as she had answered what she understood to be my last question, she turned back to some thank-you notes she was working on, no longer acknowledging my presence. I was still right beside her and had not yet even begun to get out of my seat, but she did not look my way again. I think she was that glad to be done with this latest campaign obligation, this hideous duty to share private feelings with reporters and open herself to a kind of scrutiny that might not be kind. She was desperate to move on.

She was, in many senses, the perfect foil for her husband, and the perfect ballast. He could take up so much of the room precisely because she demanded so little of it, and he could pursue even those ambitions, like the presidency, that mattered so little to her, because she did not have conflicting or competing ones. ("I have the best wife for the line of work that I'm in," Bush once said. "She doesn't try to steal the limelight.") Early in their relationship, at an extended family gathering of the Bushes, someone had asked her what she did in life and, according to Barbara Bush, Laura said, "I read, I smoke and I admire." During the campaign, Laura told me that this story was apocryphal. But Barbara told me that she remembered it well, and added that her daughter-in-law adhered to "a great philosophy in life—you can either like it or not, so you might as well like it."

The Texas governor repeatedly referred to the sacrifice that his wife was making by blessing and participating in his quest for the presidency, but when I asked him once about the biggest sacrifice he had ever made for her, all he could think to say was that he had been a devoted husband. When I asked her a similar question, she said that he would never stand in the way of anything she wanted to do, but could not cite a specific example. These were the terms of their relationship: He followed his dreams, and she followed along.

And yet there was also another dynamic to their relationship that was arguably more important and that illustrated a key ingredient of George W. Bush's luck, good instincts or talent: he chose people to accompany him through his life and career who invariably made him better and set him straight. By all accounts, Laura was more influential than anyone else in coaxing him to quit drinking. And years later, during the campaign, she, more than anyone else, kept his cockiness in check. "Rein it in, Bubba," she once said to him on the plane, as he carried on for just a bit too long. "Bushieeeee!" she would reprimand him whenever he got too testy or too silly, and those two syllables alone were enough to do it. He knew exactly what she meant.

When I joined her and Barbara in the bus that was taking them between two Michigan cities during the "W. Stands for Women" tour in late October of 2000, she was warmer again, more easygoing. Mind you, she was no font of reflexive chatter, but she waded into a subject that she and her husband usually treated as off-limits: his relationship with his twin daughters, Barbara and Jenna, who were eighteen at the time.

Every so often, George W. Bush would murmur an offhand, wistful aside about his regret that Barbara, named for his mother, and Jenna, named for his wife's mother, did not demonstrate more pride in his campaign. He let it be known that they were miffed about the disruption to their lives and dreaded the thought of Secret Service protection throughout their college years, a certain fate if he won the election. He let it be known that while he and Laura fought all the time to preserve the girls' privacy, he would nonetheless be happy for them to show up on the campaign trail more often than they did, which was almost never. They just weren't all that interested, he occasionally explained, and whenever he did so, there was a hurt look in his eyes.

On "W. Stands for Women" tour, however, Laura brought good news from this front. She told me that Barbara, who had just begun her freshman year at Yale, and Jenna, who was at the University of Texas in Austin, were tuning in eagerly and religiously to the Bush-Gore debates, even watching the telecasts with friends. They had taken offense, she said, during one debate in which Gore digressed to note that every one of his children had taken the trouble to be there, in the audience. The Bush girls had considered this a slap at them, and told Laura so. More importantly, they told their father that they were paying attention, that they were keeping up with what he was doing. Laura recalled that after one of the debates, Jenna, the more easily embarrassed and less easily impressed of the two, gave him a simple, four-word review that made him cry.

"Dad," Jenna had said, "you did great."

CHAPTER EIGHT

THE TEARS OVER THE WORDS OF ENCOURAGEment from one of his daughters weren't the only ones that Bush shed during the campaign. Behind the quips and beneath the bluster, there was a part of him that was not trapped back in his fraternity days or scented self-consciously with the macho minimalism of West Texas, that was more *mensch* than musk, and when he connected with it, he tended to mist up. Bush cried when he spoke at the Southern Methodist University event in honor of Laura in October of 1999, and he cried months later when he told an audience in North Carolina how much she meant to him, how much she had sacrificed for him. He cried on about half of the occasions when he talked about how supportive his father had been. He had a soulful, deeply emotional streak, and it grew more pronounced as the campaign intensified and the pressure mounted: Bush actually got a little deeper as just about everyone around him, including those of us in the news media, got a little shallower. And perhaps this explained one of the oddest circumstances in which I saw him cry, or almost cry: an interview that I and Alison Mitchell, a *Times* colleague who also covered the campaign, did with him during the tense, fraught period of the campaign between the conventions and Election Day.

The polls had tightened and it was unclear what would happen to Bush, and while this sometimes made him peevish or sullen, it just as often made him reflective, as he was on this day. He sat just a few inches from us, on the other side of a table in the front of his plane, but he seemed to be miles away, staring back at himself, at the campaign—at all of it—from the distance he needed to get through it and stay steady.

"It's been a great journey," he told us, a quasi-valedictory assessment that seemed divorced from his ultimate success or failure and that he rendered in a placid, resigned voice. "I'm going to look back on these days." He said he had reconciled himself to "the pressures—and, you know, I've got great faith in the American voter. I do. I've got faith in the voters."

With each passing minute of the interview, his demeanor grew even dreamier, as if he were talking more to himself than to anyone else. He made less-frequent eye contact with us. Mulling the possibility that the voters in whom he had such faith might choose Gore, Bush said, "Seriously, I would respect that. I'm not going to like it. But this is democracy." It was at this point that an accumulation of tears in his eyes became evident, and no yawn had triggered them. They seemed to come directly from the words he was saying, the thoughts he was thinking.

"I love the system and I love the country," he continued. "I love what America stands for. I don't want to sound Pollyanna-ish about it, but I do. This is an important country and an important process. And it's long and it's exciting and you've got to grip all of that: hours on the plane, policy. It is an important part, an important part of what this country's about. I am so honored to be one of two coming down the stretch. I am. And I think Americans can see my enthusiasm for what this country stands for."

He finally paused long enough to be asked: "Why are you getting teary?"

"I'm not teary," he insisted, a bit defensively. "I'm tired."

But it was more than this. Although Bush's comments strayed across the same territory he had explored with me in an interview about six months earlier, on another flight, at another pivotal crossroads in his campaign, they did so in a much different tone, with a much different affect and effect. Back then, McCain had just recovered from his South Carolina defeat with a victory in the Michigan primary that gave the Bush campaign a new shock and scare, and Bush had also mulled over the big historical picture and being a player in it. Those musings, however, contained a gee-whiz optimism and muscle-flexing braggadocio that were not yet tempered by the time and tests that followed.

"The interesting thing about these primaries is that each primary season, every four years, there's a different stamp, a new character to the primaries, a new, interesting twist," Bush had said at the time. "We're writing the history now.

"The amazing thing is, I'm in the middle of it," he noted, sounding genuinely awed. "I do find it fascinating. There's a lot of aspects to my life now that are fascinating. One, of course, is the whole history. I'm a history major. We were just talking about the history being written. It's amazing to me that I'm writing the history. And you're recording it. As a matter of fact, this interview is part of an overall mosaic that will determine the history of this campaign."

The word "amazing" came and went and came back again; he could have been a teenager raving about a rock concert. The emphasis on the first-person singular—on how neat it was for *him*, for George W. Bush, to be going through this, to be the one driving history—was consistent. And then there was his cock-of-the-walk certainty, which came through when he offered an assurance—no ifs, ands or buts—that the Republican nomination and the presidency would be his. "Mark my words," he said then.

But six months later, with Mitchell and me, this oft-used phrase was absent, replaced by the gentler, calmer assertion that "I believe I'm going to be president." And its disappearance was about more than changed polls. It was about the ways—small but real—in which Bush had matured during the campaign. He had started out with only a few stray flecks of gray in his hair and now had a generous stippling, and what had happened on the outside had happened on the inside, too. He had clearly learned something about his limitations—and about how he needed to adjust for them. He had learned that he could not take so very many things for granted, victory being just one of them, and that he could not take so very many things lightly. He was not a new man, but he was a slightly different one, and maybe a slightly better one, despite all the low blows he had landed in order to stay politically alive, despite all the hyped and fraudulent emotions a presidential campaign required.

He had once been so enamored of his gut instincts and so convinced of his intrinsic charm that he considered intense preparation for the showier, more superficial rituals of campaigns to be the refuge of sissies and sycophants. This had ceased to be entirely true. In May, nearly three months before he would take the stage at the Republican National Convention, Bush began talking with Michael Gerson, his speechwriter, about the remarks he would deliver on the big night. Over the ensuing weeks, as Bush nixed one phrase, asked for another, worried over language and made sure the words wrapped comfortably around his tongue, Gerson produced eighteen drafts, with heavy assists from Hughes and Rove.

It was more attention than Bush had ever before paid to one of his speeches. He became obsessed with one paragraph that Hughes was working on and kept sending it back to her, asking for a new try and telling her, "This is your moment. This is your moment." She finally sighed: "It's getting to be a pretty long moment, Governor." Bush was still tinkering late in the game; toward the end of July, when I had breakfast with Gerson at around eight A.M. one morning at the Four Seasons Hotel in Austin and his cell phone bleated, it was Bush on the line. He was already reviewing the latest version of the text and had a question about it. By this point, he had begun practicing his delivery of it as well, and he would continue to—in front of advisers, in front of friends, at home, on the road—right up until show time.

All summer long, friends and relatives noticed how much more committed Bush seemed, especially when he showed up in Kennebunkport for an enormous party celebrating his mother's seventy-fifth birthday and then kept a distance from the revelry, disappearing for huddles with the advisers he had stowed at a nearby inn. Others in the extended Bush clan reached for tennis rackets, horseshoes, fishing poles and bicycles. George W. reached for briefing papers and the telephone. He had stuff to think about, stuff to get ready for. He had homework.

He spent the hours before his speech at the convention on August 2 in his hotel suite with few people around him. He was uncharacteristically quiet: there was no showboating about how ready he was, no crowing about how surely he would wow the crowd. Riding in a sedan to the convention hall, he maintained his silence, then began humming softly. The song, almost indiscernible, was "Go Tell It on the Mountain."

And then, when he began his remarks, which would last about fifty minutes and be among the most important of his political life, he did something humbler than usual. He seemed to look to the spot in the convention hall where Hughes was standing, like a stage mother. I couldn't see this from my vantage point, but a colleague did, and said that Hughes pressed her arms to her sides and kept her head straight and her expression serious, in what looked like little signals and cues to Bush, who could sometimes let his arms, curl and his features scrunch up. Bush must have heeded her, because if anything, he looked too earnest, too calm, too methodical.

He approached the debates against Gore with just as much seriousness. He had begun preparing for them all the way back in May, although the first one was not likely to occur until September. (All three, as it turned out, were in October.) At the beginning of this process, he sat with advisers and studied a videotape comprising snippets of Gore's previous debate performances and high points from presidential debates in the past. By early June, he was already facing off against Senator Judd Gregg of New Hampshire, the designated stand-in for Gore, at locations all around Austin. He practiced at the governor's mansion, at an auditorium at the Texas Bar Association, at local churches and even at the studio where the actress Sandra Bullock, whose musician boyfriend lived in town, had some of her photographs taken. He practiced under bright lights and he practiced past his bedtime, which was usually ten P.M., so that he would be ready for debates that lasted this late. He practiced fielding some of the questions he least wanted to. An aide, for instance, would flash television footage of Bush hedging in public statements about the Confederate flag in South

Carolina and then press him to justify himself. Sometimes, he was simply terrible, and Hughes had to abort the session, because he was falling prey to anger and bad habits. Sometimes, he was actually good.

The skills that led to great debating were not ones that Bush naturally possessed, and his three subsequent debate performances made this clear. By any objective analysis, Bush was at best mediocre in the first debate, in Boston, and in the third debate, in St. Louis. In the second, in Winston-Salem, North Carolina, he was better than that, but only to a point. In all of them, he was vague. A stutter sometimes crept into his words. An eerie blankness occasionally spread across his features. He made a few ridiculous statements, trying so hard during one exchange to paint himself as the anti-bureaucracy candidate that he called "insurance," as in those policies that make sure we can afford necessary medical treatments, a "Washington term," as in an example of unnecessary red tape. I remember watching the first debate from one of the seats inside the auditorium and thinking that Bush was in the process of losing the presidency.

I had not accounted for the way the debate was playing on television, which was how all but a handful of voters—and, in fact, most reporters—were seeing it. And what these people noticed, because of close-ups and camera shots that sometimes focused on one candidate's reactions while the other talked, was the color of Gore's face and the quality of his exhalations. He was a pumpkin-headed sigh master, ill-served by both his cosmetologist and his petulance, and this rendered the questionable sturdiness of Bush's answers and demeanor almost wholly moot. Gore won the debate for Bush.

Ditto for the third debate, when the candidates were placed on stools and allowed to wander the stage. Gore loped and charged this way and that, even creeping up behind Bush to stare him down. And Bush, sensing that this would turn off many voters, ceded Gore the space and the energy, giving him enough rope to tie a noose for himself. Bush couldn't compete with all that rangy verbosity, so he didn't bother to try. He let Gore corner the market on authoritative. He made do with avuncular.

And he got grades from the news media that he didn't wholly deserve. It was surreal: reporters had spent the weeks before the debates writing and talking about how assiduously Bush and his advisers were trying to lower expectations for his performance. Reporters had noted how frequently Bush and his advisers mentioned Gore's vaster experience with debates, how often they alluded to an *Atlantic Monthly* cover story about Gore's evolution as a debater that featured a cover illustration of Gore as vampire, with a taste and talent for drawing blood. Reporters had made fun of all of this. And then, when Bush managed not to invent new verbs and countries or to run and cower when Gore started lobbing statistics at him, journalists pronounced Bush's performances successful, because he had *exceeded expectations*. The news media had ridiculed the game, only to join it.

"George W. Bush may well have won last night's debate," wrote Mark Sandalow in the *San Francisco Chronicle,* "by not losing it." The *Los Angeles Times* editorialized: "Bush had the advantage of lower expectations . . . mainly, Bush had to show the nation he could properly put one word in front of the other. He did that." Bully for him. Shame on us.

In our defense, there were no nights as professionally frazzling as debate nights. Wherever the event was, there would be a nearby room—usually a gymnasium or other suitably enormous space—filled with enough long tables, folding chairs, power outlets and temporary phone lines to accommodate two to three hundred journalists, all side by side, row by row, watching one of a dozen or so television sets. (The debate hall could never accommodate many reporters and would not allow laptops, and our deadlines meant that we needed to type as the candidates talked.) A few of these TVs would invariably be tuned to a different channel than the rest—and be three seconds behind or ahead of the other broadcasts—creating a most pleasing and fortuitous echo-chamber effect that was a real boon to clear thinking. We would have been better off in our hotel rooms. But then we would have missed the fusillade of documents dropped beside our laptops by Gore and Bush aides, who came along every five minutes with extremely subtle material, along the lines of "Gore is a big fat liar" or "Bush can't tie his own shoes," that reflected on something that had been said only seconds earlier. Their turnaround time was breathtaking. Their zeal was absolutely terrifying.

We also would have missed the singular pleasure, during the precious forty-five minutes between the end of the nighttime debate and our unyielding deadline for the next morning's newspapers, of having a candidate's supporters accost us and tell us how well the candidate did, an assessment that would have been offered even if the candidate had drooled copiously, cursed promiscuously and claimed to have won an early and decisive straw poll aboard a spaceship crawling with mantis-shaped creatures. Trust me: you haven't lived until you're four hundred words into a thousand-word story that's due in ten minutes, your editor won't stop calling with helpful suggestions that unhelpfully consume the dwindling seconds, you see Andrew Cuomo charging toward you and you realize that you have neither a fly swatter, a polo mallet nor a crucifix at hand. Why any reporter would waste fifteen seconds of air time or one paragraph of ink on the post-debate comments of a partisan with a predetermined allegiance and reaction was beyond me. But reporters did, so the candidates made sure to flood the press area with high-profile politicians. At the Boston debate, the politicians' handlers actually held up picket-style signs so that interested parties could know the exact spot where the U.S. labor secretary Alexis Herman, a Democrat on hand to plug Gore, or New York governor George Pataki, a Republican out to boost Bush, was lurking.

What oddballs we had all become. What freaks. I sometimes looked at myself and my colleagues and didn't recognize us anymore, so thoroughly were we enveloped in our altered states. Our immersion could be tracked by an arc of ever more elaborate technology. In the beginning, only a few reporters had hit the trail with PalmPilots. But within months, the little gadgets were everywhere, and you would see reporters on the plane or in the bus raptly beaming information back and forth to each other, as if locked in an act of technological coitus. In the beginning, everyone had hit the trail with a standard, unadorned cell phone, which rang like a phone was supposed to ring and was used like a phone was supposed to be used, snug against the ear and lips. But then one or two reporters acquired little headsets, and within a few weeks everyone had them. Reporters began setting their phones to ring with distinctive melodies or chirps, so that a bus trip could easily become a rolling, jangling symphony—a bit of "Yankee Doodle Dandy," which was interrupted by a snippet of the "William Tell Overture," which was drowned out by an otherworldly trill-gong that evoked underwater sonar—until sound itself was ruled déclassé, and the piece of equipment du jour became a special battery that made the phone vibrate and move whenever a call came in. Almost as popular were the colored face plates that turned otherwise drab gray or black phones into Day-Glo wonders.

I can still see John Berman's orange-faced Nokia, with its

vibrating battery, skitter across the table at Logan's, a sports bar in the center of Austin, on the weekend afternoons when we went to watch the pro football games there. Bush usually took Sundays off, even in the most crucial stretches of the race, and we were determined to make the most of it. It wasn't that we weren't working. Berman and I would both bring our cell phones into the bar, and we'd take and make calls, scribbling in our notebooks as fellow sports enthusiasts stared at us and as our sources, on the other end of the line, wondered what all the cheering and hollering in the background was. Once or twice, when I had a story to file, I even took out my laptop, found a space for it between the beers and tried hard to keep the blue cheese dip for the Buffalo wings and the ketchup for the French fries from dribbling down into the keyboard. (Earlier in the campaign, I had spilled some Jack Daniel's onto the "o," "p" and "l" keys, and they stopped working for the next six hours.)

It seemed that the less of a life we really had on the trail, the more attached we became to the few non-work rituals available to us and the more inviolate we declared them to be. We had, for example, a Sunday-night *Sex and the City* club, and anywhere from three to seven of us would crowd into a hotel room to see what erotic adventures Carrie Bradshaw would have this week. We were covering a presidential election, and we were living for and through a chain-smoking party girl in Manolo Blahniks. It was probably the chain-smoking we related to. (It certainly wasn't the sex.) Just about any reporter, producer or on-air correspondent who had ever smoked a little or even thought about smoking a little was now smoking a lot, although none of us really admitted it, which was why we constantly booked ourselves into non-smoking hotel rooms and then furiously concealed the evidence of our transgressions before we checked out—flushing butts down the toilet, wiping ashes off the nightstand, spraying perfume into the air and onto the drapes. Our lust for nicotine almost cost us more than the price of the cigarettes. Twice, a reporter who had not adequately concealed her crime was threatened with $100 fines for befouling a smokeless hotel room. On three occasions, several of us lit up only a few yards from the campaign plane—we had just been in the air, and deprived, for two or three hours—and were threatened by airport officials with $10,000 penalties. It was apparently not a good idea to strike matches or play with any kind of fire on a tarmac, around all that jet fuel. Who knew?

Many months later, when it was all over, one of my campaign-trail smoking companions sent me the following e-mail, a sort of summary of our experience, "I was watching the *Sopranos* last night and saw the preview for the premiere of *Sex and the City* and it made me a bit wistful," he wrote. "I think you should have a party. You should rent a non-smoking room at sub-par hotel somewhere here in town. We can all pretend to be in Pennsylvania or Michigan and order cheese sticks, over-cooked burgers and soggy fries and smoke cartons of cigarettes. The next morning at 5:30 A.M., we can all report bleary-eyed to the hotel restaurant and discuss off-season character development and whether Carrie and company can really remain friends with their lives taking them in such seemingly different directions." He was only kidding. I think.

Away from our hotel rooms, in the senior-citizen centers and high-school gymnasiums and outdoor football stadiums where Bush sought to close the deal with American voters, the candidate himself did an intermittently good job of keeping us entertained. In one speech, he defined the importance of stable homes by saying that "families is where our nation finds hope, where wings take dream." In another, he extolled "people who perform common, common acts of, commonplace miracle—acts of miracle every day." And in yet another, he departed from the TelePrompTer text he was reading to note that he did not merely believe that salmon-friendly hydroelectric dams could be built in the Pacific Northwest but that, beyond this, "The human being and the fish can co-exist peacefully." It was much-needed succor for readers of Herman Melville and Peter Benchley and for the many Americans worried about the possibility of a school of red snapper pointing a nuclear missile at us.

He was puffed up and pumped up by his disaster-free navigation of the Straits of Debates. His advisers were positively ecstatic, and they were never very good about keeping their emotions under wraps. After the second debate, in Winston-Salem, a television news producer overheard Hughes on the phone with Johndroe, who was being instructed to relay a message to Bush. The message was not simply that the instant poll numbers, gauging viewer reaction to the debate, looked spectacularly good for Bush. It was, as Hughes rendered it, that he "just became the president of the United States."

This was the campaign's mood when, five days before the election, a story with potentially devastating consequences broke. A Maine television station reported that in 1976, when Bush was thirty, he had been arrested in that state and pleaded guilty to a misdemeanor charge of driving while intoxicated. In all of Bush's discussions about his reckless youth and in all the interviews in which he had been asked about his past, he had never mentioned this, a decision that he and his advisers had very deliberately made. Aides said that Bush had not talked about it because of the example it would set for his daughters and other young people: How could he convince them that such behavior was destructive if he had indulged in it and ended up where he did? This was as reasonable an argument as Bush or his advisers were likely to muster. Less easy to digest was the way they reacted publicly to the discovery of his secret.

Bush held his first news conference in about a month, acknowledged the arrest and stated emphatically that he was not proud of it. But he then, quickly and defiantly, proceeded to plant suggestions that the emergence of this information was a matter of political hanky-panky. He spent as much time and as many words being aggrieved by this possibility as he did being sorry about what he had done and what he had concealed. This, to me, lifted the story up a few notches, compelled us to pay more attention than we might otherwise have. Bush should have been mildly ashamed, and was instead acting like a victim.

But he caught a big, big break. News organizations don't like to publish or air incendiary, potentially prejudicial stories about candidates in the final days; they don't want to be seen as, or accused of, trying to tip the election one way or another. So they hurry to run all of their most critical dissections of candidates' records or personal histories well before the last minute, and if something like Bush's drunken-driving arrest pops up in the home stretch, well, they cringe. They pause. And, sometimes, they end up giving the candidate more of a pass than he or she might otherwise receive.

My first instinct was to mention the revelation of Bush's arrest—and his explanation of it at the news conference that he and his advisers, finding the situation serious enough, hastily called—at the very top of my story for the next day's paper. I felt that Bush had known people would look into his past, had known he had an arrest *on the books,* and had never said anything—all of which made the situation about something larger than just how he had behaved at the age of thirty. With his news conference, Bush himself had made the past arrest something current by addressing it head-on, by ending one of his last days on the trail with a tense encounter with the press. But the story that I ultimately wrote, in consultation with editors, played all of this in a tempered way, leading with several paragraphs about Bush lashing out at Gore for what Bush maintained was an overly bureaucratic vision of the federal government. Readers had to stay with the story beyond that point to find the first reference to the arrest.

Readers also got a truncated version of the strange fallout the next day. On the plane, Wayne Slater of the *Dallas Morning News,* who had covered Bush in Texas before the campaign, recalled a conversation he had once had with Bush in which he had asked the Texas governor whether, after an arrest in college for a prank, he had ever been arrested again. According to Slater, Bush had said no and the interview had moved on. But then, Slater remembered, Bush seemed to indicate that he was about to circle back to the question he had just answered and maybe even amend his response. At that point, Slater said, Hughes interrupted and more or less stopped whatever was about to happen. As Slater went over all of this for a huddle of reporters in the middle of the coach section of the plane, Hughes nervously offered her own version of events to another huddle of reporters just a few rows away. She said that neither she nor the governor recalled responding to that question in a negative and false fashion, but she said that she remembered feeling that Slater had emerged from that interview with an accurate impression. An accurate impression of an unacknowledged event? It was an awfully weird concept, and Bush and his advisers didn't end up taking as much heat for it as they perhaps deserved.

If this whole sequence of events surrounding the news of Bush's arrest had happened, say, a month earlier, reporters would have paid it more attention. I am not saying we would have made it the story of the century; there were limits to its relevance, and it hardly disqualified Bush from the presidency or refracted his entire life and character through a different prism. But we would have treated it with less trepidation, and I don't think Bush understood or appreciated this. I could hear him, on those final days on the campaign plane, mutter to aides that the story had gone away because the voters smelled something underhanded behind it and because they saw his past arrest as no big deal. Maybe they did and maybe they didn't, but this was most certainly not the full explanation for why things turned out the way they did.

His aides were especially grateful that the *Times* had shown restraint, because they felt that the major television networks took their cues from the *Times* more regularly than they did from anything else. This was why, on the second morning after the story broke, Bush sauntered toward me as I cadged a bagel from the refreshment table between first class and coach and said, most originally, "You're a good man." I tried to take advantage of it.

"In that case," I said, "I have an interview request pending. Why don't you give me an interview?"

At the mention of the word "interview," Hughes, seated just a few rows ahead of where we were standing, bolted upright and sped toward us. She shot me a disapproving glance; I knew, and she knew I knew, that interview requests were to be negotiated with her, not with Bush. (I never did get this particular interview.)

I smiled. "You can't blame me," I told her. "He complimented me, and I had to seize the window."

It was an awkward, stupid phrase, and Bush realized it before I did.

"'Seize the window'?" he said, a grin consuming the entire lower half of his face. "You're talking like me!"

This was the thing about Bush—whenever you wanted to dismiss him as slow-witted or unreasonably pleased with himself, he would do something along these lines. He would show, in an instantaneous response, a flash of cleverness and a clear self-knowledge about his own failings. And he would be likable, just as you were dwelling on aspects of him that weren't so likable at all. I don't think any of this was accidental. I think that on some instinctive level, Bush sensed little disturbances in the atmosphere around him and calibrated his actions accordingly. Politicians are seducers—at least the good ones are—and Bush was practiced in the art of seduction.

As much as he had adjusted to and grudgingly accepted the tedium of the trail, his mood grew visibly brighter over the last three days; he could see the light, and he was hurtling toward it. He was also being told by his political advisers, most of whom shared Bush's tendency toward overconfidence, that the election was more or less in the bag. Rove was predicting 320 electoral votes—not a landslide, but a comfortable majority, since only 270 electoral votes were needed to win. He was predicting a four-to seven-point victory in the popular vote. And Bush was beaming.

He had long razzed Gore for Gore's past claim to have had a role in inventing the Internet, a sloppy statement with a true subtext: Gore had indeed promoted legislation that helped the Internet along. Now Bush was asking crowds why, if Gore was so smart, all addresses on the Internet began with "three *W*s," Bush's middle initial. It was a moderately cute riff the first time Bush tried it. It was sophomoric the third and fourth time. And by the ninth or tenth, it was actually kind of painful, especially when Bush translated it into Spanish for an audience in Florida. ("Tres doble *V*s!" he trumpeted.)

"Can't you make him stop?" I asked one of his advisers.

"No," the adviser said. "He loves it. *He* came up with it." And this, of course, wasn't true of most of Bush's signature lines.

With seventy-two hours to go, the rallies waxed grander and nuttier. In Dearborn, Michigan, on November 4, Senator Ben Nighthorse-Campbell, a Colorado Republican, took the microphone and sought to convince Bush supporters of the importance of voting by saying that Adolf Hitler had been lofted to power by a single vote. It was an analogy that even Gore's most vitriolic detractors must have found out of whack. A local Teamsters official who introduced Bush and several of his political allies inadvertently referred to General Colin Powell as Adam Clayton Powell Jr., a black Democratic congressman from New York City who had died in 1972.

Bo Derek was also on hand; she was starting to pop up here, there and everywhere that Bush went. To this day, I have not received a satisfactory explanation for the wisdom of showcasing Derek; at this rally, for example, Bush was going to talk about traditional values, which didn't gibe with the presence of an actress who had posed nude for *Playboy* and moaned her way through soft-core flicks. But here she was, making reference to the one quasi-gem in her movie oeuvre. She implored each of the people in the crowd to get "10" or more friends to vote for Bush. "I've done it in Hollywood," Derek said, leaving the "it" rather ambiguously defined. "If I can do it in Hollywood, you have no excuse."

The Dearborn rally at which Bo appeared was indicative of the kinds of bells and whistles, no matter how nonsensical, the candidates were attaching to their campaigns in the final hours. They were desperate to make big gestures. Gore committed himself to an almost sleepless, thirty-hour marathon of campaigning until the polls opened on the morning of November 7, while Bush devised a schedule for November 6 that began in Tennessee and ended in Arkansas, the states that had given political birth, respectively, to Gore and to Clinton.

Bush flew into Bentonville, Arkansas, his final stop on the way back to Austin, well after the evening television news shows and the first-edition deadlines for daily newspapers just about everywhere except the West Coast. All day long, the campaign's advertising chief, McKinnon, had promised us a musical surprise, but we were skeptical. The legends of song who showed up at Bush's rallies tended to be fossils like Loretta Lynn and the Bellamy Brothers or cornpone crooners like Billy Ray Cyrus and Travis Tritt. The plane landed and we all filed into a hangar at the Bentonville airport and looked toward the stage. We spotted the über–lounge lizard of Las Vegas, Wayne Newton. Was *this* the special entertainment? We spotted Derek. Was she going to burst into "America, the Beautiful"?

As Bush made his way toward the microphone, a recording began to play. No, wait, it couldn't be. Yes, yes, it was—Fleetwood Mac's "Don't Stop (Thinking About Tomorrow)," which Clinton had used as his theme during the 1992 campaign that unseated Poppy. OK, so the son, here in Clinton's home state, was appropriating it. What was so ingenious about that? Then, smack in the middle of the song, it stopped, the moment marked by a scratchy sound that evoked a needle being lifted indelicately from a record on a turntable. And a new song, or rather the climax of the chorus of a new song, took its place. "We won't get fooled again!" bellowed The Who. Not bad. Not bad at all.

Naturally, Bush, the pop-culture alien, had had nothing to do with it. He told us back on the plane that he hadn't even known it was going to happen. He looked wiped out and blissed out—relieved, beyond words, to be done with it, to be heading home.

Next stop, Austin, where a home crowd and his own bed awaited him. His posture uncoiled. Every bit of tension in his face dissipated. He once again looked more childlike, but what didn't go away was the grown-up perspective he had acquired, his palpable feeling that this was really pretty serious stuff and his discovery of a respect for it deeper than he had once let himself admit or show.

He got on the public address system. He had done this on the way to Bentonville—to note that it was our last chance to hear our favorite bungled phrases of his—and now he was doing it on the way to Texas, with a more emotional message. "I do want to say, on our last leg, how much I appreciated most of you," Bush began. "Well, maybe all of you. I do appreciate the amount of time you've taken away from your families, how hard you've worked, how hard you've been professional in this campaign.

"I didn't particularly care for everything you wrote or said on the air, but I do respect you as people," he continued. "May God bless you and your families. Thank you for your hard work for America, and make sure you vote for me. Good-bye." He then walked slowly down the aisle, shaking each and every person's hand. When he returned to his seat, he instructed the flight attendants to play a videotape of the *Saturday Night Live* skits that had ridiculed his debate performances as vacuous and Gore's as overbearing. It flickered across all of the television monitors in the plane, which soon filled with laughter. Bush seemed just fine with this.

CHAPTER NINE

IT WAS ONLY ABOUT FIVE MINUTES AFTER THE network newscasts augured a Gore victory in Florida that I and the other reporters in the small "media pool" on Election Night were herded out of the Four Seasons hotel in Austin and onto a bus. There was no reason to stay. Bush had fled. He had been dining in an adjacent restaurant with his wife, daughters, parents, siblings and other relatives, after which he was supposed to join a big group of friends to watch the returns in a suite at the hotel, perched above Town Lake. But Bush wasn't through dessert when he decided to junk it all, to head home to the governor's mansion. We followed. If we were going to be allowed to see him, it would be there.

The campaign aides with us were grim-faced and silent, not even bothering to explain why Bush had departed so suddenly and unexpectedly. They didn't need to. At 6:10 P.M. Central Standard

Time, when he and his family members had arrived at the Shoreline Grill, it already looked like two other pivotal battlegrounds, Pennsylvania and Michigan, were going to Gore. The disappointment of this and the nervousness it created were visible in the Bushes' gaits and expressions as we watched them climb from dark sedans and go into the restaurant. The Texas governor could see a bunch of us standing just yards away, and someone even shouted a question at him, but he didn't have any words, smiles or waves for us. He scooted out of the public eye and into the restaurant as fast as he could.

The news about Florida, announced about forty-five minutes into his meal, seemed to seal his defeat, and he bolted for the private, quiet confines of home. There were subsequent reports from people inside the restaurant that, before he left the dinner table, Jeb, who was with him, wept and apologized to him. There were reports that the two brothers bickered, heatedly enough that their father had to intervene, and there were contradictory reports that George W. simply hugged Jeb, who was in tears, and told him it was all OK. The family members—the only people who know for certain—have never talked about this in any detail. But whatever happened, Bush's unplanned exodus made this much clear: If he was in the process of losing the election, which it certainly looked like he was, he didn't want to do it in crowded company—with so many eyes upon him, with so much pressure to be stoic. Only his parents and his wife accompanied him back to the mansion. Not even Jeb followed them, at least not at first.

His aides didn't know what to do with us: we were supposed to keep the "body watch" on Bush for the rest of the news media and to be brought in to get a glimpse of—and some comments from—him as the returns came in. This was why we had been stationed in a ground-floor meeting room at the Four Seasons, but where were we to be stowed now? The aides put us on the bus to get us closer to Bush and the mansion while they figured it out. It was craziness: no one knew what to do, and nothing was going according to plan. As the bus weaved through downtown streets bedecked for a victory celebration that would never come to pass, I speed-dialed colleagues on my cell phone, letting them know about Bush's sudden change of heart, letting them know how significant it felt. The night had turned weird—I knew this much even then. What I didn't know was just how weird it had turned.

Along the way to the mansion, CBS News volunteered to give the members of the media pool seats and a bit of work space in one of their rented trailers, about a block and a half from the mansion. This way, we would be nearby if Bush decided to let us see him. He did—about ninety minutes later. We marched into the mansion, about twelve of us, and upstairs to the private living room, where Bush, Laura and his parents sat around a fireplace, a television set nearby. By this point, Bush's senior advisers had determined that Florida was still an open question, that the declaration of a Gore victory was premature, and Bush wanted to get that message to voters in places in the country where the polls hadn't closed. He tried to look upbeat. He more or less succeeded. I asked him how a person got through a night like this, with his whole future on the line.

"Actually, my whole future isn't on the line," he said, a reminder of all those times he had assured us that he had not invested and would not invest every bit of himself into this single ambition, this one goal. Then he added, "I'm not worried about *me* getting through it." With his head, he motioned to his parents.

Barbara Bush actually looked fine—perhaps a little put out by the uncertainty and the waiting, but by no means despondent. "We haven't been up this late in years," she said, as if the main point of suspense was when they could all get to bed.

The former president, however, looked like he was trying to smile through an evisceration. Asked to compare this night to the same one eight years ago, in 1992, when he lost to Clinton, he said without hesitation that what he was going through now was a "hell of a lot worse." His haggard expression and empty gaze backed up his words. Whether he was in the grip of simple paternal devotion or a more complicated concern for his and his family's legacy, the elder Bush was clearly being torn apart by it.

The Texas governor's face was braver, and more tranquil, which was something of a riddle, given his retreat to the mansion, an apparent signal of panic and depression. Over the course of the campaign, Bush had clearly learned something about composure, and how to regain it quickly. He had learned to press his talent for detachment into the service of a useful and reassuring equanimity, something that would prove to be even more helpful to him down the line, during the most challenging days of his presidency. He leaned on his good, friendly humor in a similar fashion, and had done so the minute we were ushered into his private living room in the mansion that night. When he had started speaking, stating his continued optimism about the election's outcome, he noticed that I had a worried expression on my face and was punching buttons on my tape recorder in frustration. He stopped what he was saying, even as the television cameras rolled, to find out what was wrong, and I pointed to the uncooperative little machine in my hands.

"My man Bruni is experiencing technical difficulties," he said in an amused tone of voice, with a matching smile, then waited for me to correct them before he started over again, from the top. Bush could be that way—a kind and gracious sport under nail-biting circumstances.

Jeb wasn't with his father and older brother when we were in the room, but a short while later, after we left, he joined them. The night was visibly harder on him than anybody else. For weeks leading up to it, as the polls in Florida showed a race there tighter than the Bush campaign had ever thought it would be, political operatives and journalists in the state openly questioned whether Jeb had been devoted enough to his brother's cause. At rallies beside his brother during that final stretch, Jeb's manner was wholly different than it had been months earlier—more ardent, more frantic. He would stand at the microphone to introduce George W. and state his love for him more emphatically than before, and he would beseech the voters within the reach of his voice to help him keep his promise, the promise that he would put Florida in the Bush column. He was as nakedly emotional as I'd ever seen a politician onstage, and it suggested just how many of the people closest to him he felt he would be letting down if Florida fell. And on Election Night, he was watching this happen.

In the upstairs living room at the mansion, he monopolized the phone. He called Republican precinct captains throughout Florida to ask them for the latest tallies, the latest sense of things, county by county. Laura Bush had drifted off elsewhere in the mansion for a while, and Barbara Bush was at the edge of the room, several yards away, pacing the wooden floor, her hands clasped behind her, her mouth uncharacteristically shut. The only ones huddled together, at least for a time, were the three men, the former and current officeholders, the bulwark of the dynasty. As Jeb rounded up results, his father weighed in with thoughts about the demographics and political dynamics of the various counties Jeb was checking on—this was the family business, and the former president was still immersed in it. Quietest of all was George W., the one whose fate was in the balance.

Over the ensuing weeks, I would come to wonder if that reticence and stillness were about more and less than simple steadiness, if Bush was going through a fresh bout of ambivalence, an old ghost come back to haunt him. He had agreed, in a happy-go-lucky fashion, to run for president, and he had proceeded, in a more judiciously happy-go-lucky fashion, to make a decent go of it. He had devoted time and attention to the obligations of campaigning that he found most irrelevant and ridiculous. He had come to treat the task that he had set for himself with more deference and respect, or at least to communicate in his public bearing that those emotions were inside him. He had even seemed to sharpen his desire for the prize.

But was *this* the prize he had bargained for? A victory, if he managed it, so narrow and ambiguous that it would send him to Washington without a mandate or heady sense of triumph? An outcome guaranteed to heighten partisan acrimony in the capital? For all of Bush's efforts to make his trek toward the White House as agreeable as possible, it was now ending in the most disagreeable manner possible—in a stalemate that would go on for thirty-six days, a surreal political and legal limbo that would live on in history books. Watching him in and around Austin over the period between Election Day and the Supreme Court judgment that effectively handed him the presidency, I could not stop wondering if he was having second thoughts.

On Election Night, after I had seen Bush for the last time and the pool was disbanded, I went to an outdoor tent near the Texas State Capitol in which tables, electrical outlets and phone lines had been set up for the scores of journalists in town. The temperature dropped below 50 degrees, cold enough that I could barely punch the keys in my laptop. I was supposed to write a mood piece to go along with the news story of Bush's victory or defeat, and I kept redoing the top, tilting it this way or that, getting it ready for whichever verdict came down. Whenever it eventually did, I and all the other reporters around me would have just minutes to file our stories. We waited and shivered and watched our television sets just like everybody else, until it was past midnight and then past one A.M. and then past two A.M. I had cadged all of four hours of sleep the previous night and felt like the walking and typing dead. I guzzled coffee. I popped ibuprofen. And at three A.M., I packed it in, wondering where this would all go from here.

The next four weeks were the strangest reporting experience of my life, because I was in the middle of the biggest story imaginable and had almost nothing to do. The action moved to Tallahassee, the fight was in the Florida courts and Bush and his advisers didn't quite know how to handle affairs in Austin. So they played it quiet and cautious. Every few days, they would activate a small pool of reporters to drop in on Bush as he met with future administration officials or sat in a chair at the mansion or walked to or from a car. That was it. Most of his advisers stopped returning phone calls, because they didn't want what was going on to seem political, and they were the political hands. They sought to portray Bush as a patient leader awaiting a fate being decided elsewhere, by distant judges, and not as a desperate competitor trying to affect the result. That was surely part of the explanation for the way he appeared in public—hesitant, halting, fatalistic—on those infrequent occasions when he did appear. But it was perhaps not the full explanation.

When he was in Austin, he stayed close to home and never missed his late-morning or midday workout at the University of Texas fitness center. But he wasn't in Austin all that much. He spent more time about 90 miles northwest, at his ranch near Waco, where he did not then have cable or satellite television, just a few local channels. Much of the rest of America was glued to

CNN or MSNBC, with the networks' minute-by-minute updates of a story that had no meaningful updates. Bush was running, reading a Joe DiMaggio biography and, we were told, doing manual labor on his 1,600 acres. He was clearing the brush from the mouth of a canyon—bushwhacking, if you will. He was cleaning up the tangles that he could clean up, after withdrawing to a world that obeyed more predictable rules and expectations.

"More than anything," one of his longtime friends told me in a telephone conversation one day, "he's uncomfortable. He doesn't like the mess." This was a family trait: the Bushes, true to their stock, did not like conflict or controversy. It was why his father had never mucked around much in social issues and was able to lurch sharply to the right in his stated thinking about abortion and wonder why everyone made such a fuss about it. Government, to his thinking, wasn't really about such private, personal matters. It was why his son wanted everything around him to be cordial and merry. The Texan in him put an extra emphasis and an earthy spin on the merry, but the New England WASP in him set the cordial baseline.

During those first weeks of the recount, he looked shaky and tentative whenever he broached serious topics in a serious setting. But whenever he took his tie off and stood on the dust and grass at his ranch, he was a new man. The first time a pool of reporters was taken to see him there after Election Night, he went out of his way to steer the conversation in more trivial directions, to leave the heaviness behind. Rather than ignoring the barking of Spot, who was conveniently nearby, he used it to divert attention from himself and his predicament. He chastised the spaniel. "Sorry," he told the reporters after one of Spot's interruptions. "The dog wanted to have a few comments. What she was saying was, 'Let's finish the recount.'"

He wanted it to be that simple, that easy—a desire that had always seemed antithetical to the task he had taken on when he first stepped onto the campaign trail, a desire that he had not, by all outward evidence, fully relinquished. He lingered at the ranch for a day longer than his aides said he would probably be there, and then a day longer than that. In order to conduct some Texas business that he needed to attend to, he had a carload of state officials drive up from Austin to see him. His spokeswoman in the governor's office, Linda Edwards, said that it was the first time in his nearly six years as governor that she could remember anything like that happening.

With the media invasion of Austin, rooms were hard to come by, and I switched hotels three times. Every night, the bar at the Four Seasons turned into a meeting ground for dozens of journalists, ones who had been together much too long and ones who were dropping by only now, drawn by the promise of an excitement that existed elsewhere but not here in Texas. Walter Cronkite showed up in the bar one night and approached Richard Wolffe and his wife—she had flown in from Washington with their infant daughter, Ilana, to visit Wolffe. As Cronkite leaned heavily on his walking cane and smiled at Ilana, he asked Wolffe: "Does she realize what momentous times she's living through?" Wolffe responded, "I'm trying to get her to read the papers, but she won't sit still." Andy Rooney was also in the hotel, and when he entered the bar at one point, he was overheard asking a waitress, "Do you serve liquor?"

When Bush did come back to Austin, we didn't see all that much more of him. The less he showed his face and said, the more each sighting of him and every word that escaped his lips qualified as potentially newsworthy, so camera crews staked out the back driveway of the governor's mansion and the front steps of the

State Capitol. Reporters milled around a press-briefing room that had been set up at his campaign headquarters, just in case one of his advisers decided to step up to the microphone and talk. We spent as much time calling each other's hotel rooms, wondering if anyone was getting any more information or any more cooperation than the rest of us. On my television set one morning, I saw Jeanne Meserve answer a question from the CNN anchor in Atlanta about how the Bush camp was feeling and what it was doing by holding her cell phone up to the camera and saying that she was waiting for it to ring.

Sometimes Bush's aides activated a pool to watch him walk into the State Capitol in the morning—the statesman at work—but this yielded little. He would step and smile and wave and smile and step; he could have been the Rose Bowl Queen, minus the tiara and float. This was the pool report I filed for my similarly addled colleagues on one of the days I drew pool duty, an attempt at a humorous reflection on our bondage:

> This is the pool report you've all been waiting for—brimming with incident, bursting with color, so electric and eclectic it may leave you apoplectic.
>
> Where to begin?
>
> There was a bracing chill in the late-November air around the State Capitol and the sun hung low in the east when Gov. George W. Bush loped into view at 7:51 A.M., smiling and seemingly energetic. That time, for the record, is according to my Qualcomm cell phone clock (with Bell Atlantic Mobile service). The clock is digital and gives only minutes, not seconds, hence the crudeness and sloppiness of the time estimate.

The governor first espied John Berman of ABC News. Although NBC was the official TV pool representative, Mr. Berman had gathered a crew about 20 yards from the pool's perch on the steps leading to the south entrance to the State Capitol.

The governor's first words are lodged forever in my memory.

"Berman," he said. "How are you?"

Then the governor noticed something different and disconcerting about Mr. Berman—something that hinted at the long, grueling, redundant, depressing vigil of the reporters and producers and photographers around Austin. For at least 36 and possibly 48 to 72 hours, Mr. Berman had not come anywhere near a razor. . . .

But where other observers might have seen, on Mr. Berman's cheeks and jowls, follicular decrepitude and hirsute disrepair, Mr. Bush saw something else. An optimist, he saw the seeds and stalks of new growth.

"I like the beard," he said, still loping, never breaking stride, several bodyguards around him. "Very stylish."

The governor himself was clean-shaven, and had clearly accomplished that task with the requisite agility, as there was not an adhesive bandage anywhere on his face.

He was now only 10 yards or so from the building, and he was moving fast—so fast that a complete sartorial appraisal was not possible, though I'm sure there will be TV images played all day, and these will allow the interested to deconstruct his wardrobe. What I remember was a dark suit, a shirt that was either white or light blue, a reddish tie. It was serious business dress.

At some point he said, "Doing great, thank you." The question that triggered this remark was inaudible.

At another point, when he was nearest the pool, he said, "Good morning, everybody. I miss you all." I have an actual tape recording of this part, should anyone need to assess the emotional timbre of his voice.

We were losing him. He was ebbing, receding, getting ever closer to the door.

Alexandra Pelosi shouted: "Governor, how do you think the recount is going?"

Those words floated and swirled and did little arabesques in the air for what seemed like an eternity—well, at *least* half a second—and Mr. Bush, digging deep into his soul for a response, said:

"Good to see ya."

And then he was gone.

There was no telling when it would end, and every time a final outcome seemed as close as a few days away, the imagined deadline came and went. At crucial moments, Bush went on TV, trying to reassure the country that everything was fine but coming across as squinty, squeaky, slight. After all of this time, he had still not mastered one of the basics of modern politics: looking into the lens of a camera and making the people watching him on their television sets forget that this is what he was doing. It was an unnatural kind of communication and, more often that not, he looked wholly unnatural doing it. There were certain boxes he couldn't fit himself comfortably into, no matter how many times he tried.

The suspense and uncertainty that he was experiencing finally found release after the Supreme Court's decision, when he at last got to name his first cabinet officer, holding a news conference near his ranch to introduce Colin Powell as the new secretary of state. His eyes watered up as he did it, as he let go of all he had been holding tight. Then he allowed his mind to drift elsewhere. While Powell spoke, he searched the crowd, locking eyes with reporters in the room, making idle contact with him. He later teased one of them for having worn sneakers to the event. It was a detail that Bush, the newly minted president-elect, had actually fixed on.

I was gone by then. About two weeks into December and two days before the Supreme Court ended the stalemate, my editors had finally sprung me for vacation. I turned off my television set and didn't lay eyes on Bush again for nearly a month. Then, about a week before his inauguration, my colleague David Sanger and I, who would be covering the White House together for the *Times,* were granted an interview with Bush at his ranch. We spoke with him briefly in Washington the day before, then flew to Waco and drove the remaining distance to Crawford, the tiny town of about seven hundred people that was closest to his property. We entered the main gate at about nine A.M. We didn't drive back out until noon.

The Bushes were nearly finished building a new, 4,000-square-foot house by a bass pond that had been specially dug and filled with water and fish, but they hadn't moved in yet. They were still staying in an older and smaller house a good half-mile away. Bush was standing outside, in jeans and an open-collared shirt, when we drove up. In addition to Spot, a black Scottish terrier named Barney scampered around his feet. Bush had gotten the dog as a birthday gift for Laura from New Jersey governor Christie Whitman, whom the couple had seen during a swing through New Jersey just a few days before the election. Spot and Barney followed us inside, and as we made our way to the kitchen, where Bush had brewed a fresh pot of coffee, we heard a blood-curdling squeal, turned around and saw the *Times* photographer who had come with us lifting his foot from Barney's flattened torso. The photographer had inadvertently stepped on the pup. Bush rushed over, scooped Barney up, cradled him and kissed him lavishly on the head.

The photographer, who had been to the Bush compound in Maine more than once, noticed a familiar green in the decorative scheme of this place. "Is it just me," he said, "or do these colors remind me of Kennebunkport?"

"It's just you," Bush said.

It was not an auspicious beginning.

When we moved into the family room, where a miniature sculpture of the White House hung on one wall, Bush unrolled an enormous sheet of paper that represented an aerial survey—more like a satellite photograph—of the property. He pointed out this feature and that, where the canyons were and the snaking line of the river that ran through them, and he seemed in no hurry to get to the substantive stuff. I had been here once before, in a group of about two dozen reporters to whom Bush had given a tour in July, but Sanger hadn't, so Bush treated him to the whole spiel. The president-elect could talk about the different kinds of trees and wild grass on the land for hours. I almost thought he would. But after about ten minutes, we all sat down and the interview began.

This discussion lasted nearly an hour, and ranged responsibly over various countries—Russia, China, Iraq—and various topics: tax cuts, the environment, civil rights. Bush made clear what he would later, in his first months in the presidency, make evident: that he thought the Clinton administration, especially in its final months, had gone too far with environmental regulations and land protection, especially during what Bush considered an evolving energy crisis. He shrugged off the controversies over his selections, for his cabinet, of Gale Norton for interior secretary and John Ashcroft for attorney general. Bush always seemed a bit taken aback by these predictable furors, so confident of his good judgment and good intentions, in that Bush-family way, that he never fully understood why people didn't just trust him to do right. His was the politics of character—*I'm a decent person, so why do people doubt me*—and not the politics of ideology, of doing explicit battle for one school of governance against another. But he had positions that indeed favored one ideology over another, and now that he was on the national stage, he was getting a first-hand lesson in the discord that was bound to sow.

Toward the end of our talk, I asked him about a stack of books on an end table nearby. They had titles like *White House Furniture* and *White House Art,* and were essentially decorating manuals for incoming occupants of the White House. They existed so that a new president could make his own selections of available art for different walls, available furniture for different rooms, with a helpful code to let the selector know which pieces were considered appropriate for which niches. What must it feel like to be one of the very few people in history to actually dictate the look of the nation's most hallowed residence, to supervise that surreal commingling of personal taste and national tradition, to claim 1600 Pennsylvania Avenue as your own? Bush certainly seemed to be enjoying this part of it. He leafed through the books and showed us some of his favorites, including an old painting of Washington, D.C., as seen from the naval yard at the time, that enlarged the look of the White House so that it wasn't dwarfed by the U.S. Capitol. He said he liked the statement that made about the balance of powers between the executive and legislative branches.

He had decided to use, in the Oval Office, the desk that had once been John F. Kennedy's, and we asked him what had happened to his father's. "Storage," he said, and the word hung in the air for what seemed an awfully long time. At both our prompting and his own, the subject of his father entered the conversation repeatedly, and it was clear that the family's accomplishment—a father and son, both presidents—had hit George Herbert Walker Bush and George Walker Bush full-force. In fact, when Sanger asked Bush what he was currently reading, Bush went into his bedroom to fetch the book and came back with a biography of John Quincy Adams, the only other son in American history to follow his father into the presidency.

"I might as well find out what the fellow was all about," he explained. Asked if the inquiry was proving to be instructive, he noted that Adams feared that his rise in politics would be seen as nepotism. "The dad writes letters assuring him that it's not the case," Bush said.

I noticed a painting above the fireplace of two familiar-looking men in a small boat: a fishing scene. What was it? He explained that it was a portrait he had commissioned of him and his father on one of their favorite lakes in Texas. What was most interesting was how the younger Bush had dictated the positioning of the figures: His father was in the foreground, facing the observer and in the midst of reeling in a big catch. The son was behind him, in profile, less easily noticed, no fish on his line.

We eventually left the family room to climb into several open-air, all-terrain vehicles for a Bush-guided tour, but not before he obsessed over the fact that Sanger and I were not wearing jackets and would get cold as we whipped around. He went to a closet full of stray extra clothing, and when I told him it really wasn't necessary, he said that I should just take his advice, because he didn't want to hear me gripe later on about failing to. We plucked old jackets of his from hangers and put them on, and off we went.

Bush had been in the public eye too long, and had endured too many interviews, to drop his guard entirely, but the relaxed, informal nature of the next hours brought out intriguing bits and pieces of conversation. Several times he mentioned his daughters, who had never learned to share his affection for this place, so removed from Austin and their friends. As we climbed across a bed of slippery rocks to get further into a canyon and a better view of a waterfall there, he mentioned that Jenna had consistently questioned his decision to buy and build on the property until she saw this scene, occasionally teeming with wildlife. He was clearly thrilled that he had got through to her, just as he had clearly been bothered by her aloofness. He had often said in speeches on the campaign trail that dealing with teenage daughters was no picnic. He had often hinted that they were not as impressed with him as he wanted them to be. On this morning, as he took us through the new house and pointed to their bedrooms, Sanger asked if he missed his daughters, who had moved into college housing four months earlier.

"Not really," he said. "They'll be around. Trust me. I love them, but I'm really thrilled they're out on their own." He talked to them on the phone only about once a week, he said, but that was OK: Laura talked to them much more frequently than that, and he got the reports. Most of our conversations with him over the course of the morning were like this—utterly casual, mostly deprived of any sense of moment. He wanted to talk about the little things, not the big ones. He asked after other reporters from the campaign trail who were now suddenly absent from his life, companions of a sort who had scattered. I gave him the updates: who would be covering the White House, who was forever gone, who had moved to a new city. John Berman had become engaged and Bush said that he had sent Berman a note wishing him the best. I later learned that it said, "I trust the patience you learned following me from Iowa to Maine to California and, yes, to Iowa, will make you an understanding and compassionate (and conservative?) husband."

I kept waiting for Bush, now bound for the White House, to seem ineffably grander than he had before, but he still seemed much like an ordinary man who had lived through a chapter in American politics more extraordinary than he had ever expected or counted on. At one point during our tour of the property, he noticed, in the dirt, a fossil-like piece of jawbone with jagged teeth and stooped to pick it up. He couldn't figure out the species. "Whatever it is," he said, "it's got a few cavities." He tried for smooth words and unknowingly butchered them. He said it was "dexterious" of me to make it over a rough, sloped patch of land in black loafers. He talked about being "ameniable" to certain suggestions.

He knew, from his parents, more than a bit about White House life, and he said that the perk he expected to enjoy most was "good food, I guess. They've got great food. The dessert menu is unbelievable. These things are so exotic, it's hard to even describe what they are. They're these kinds of mounds of calories that could be ice cream, could be—yesterday, at the Blair House, we had this, I'm not even sure, coffee ice cream surrounded by this unbelievable meringue, beautiful meringue, it looked like a giant pineapple. It was really good."

He imagined the fun of introducing friends to this. "I've got a lot of friends, and it should not surprise anyone that we'll be surrounded by friends," he said, referring to himself and Laura. "Our buddies will come from Texas." He mentioned one couple in particular, saying that "they will remind me—they will keep me from being so hermetically sealed that I'll stay in touch with reality. Laura will be good at that, too, by the way." He added that these friends were not "political friends. I mean good, solid friends. They were friends before I was president, and they'll be friends after I'm president. That's going to be one of the really fun things for us to do, is to see them up there."

And he said that he was, indeed, looking forward to the challenge ahead, but the conviction in his voice was fettered. He knew, and acknowledged, that he was walking onto a minefield, that the prize he had won came with complications because of what had happened in Florida, because Gore had prevailed in the popular vote. It was a diminished thing. He put the thought away and dwelled on how much he would miss Texas, how determined he was to get back here. He would return, he said, more often than people might imagine, and there was something defiant about this vow. He would bend to expectations only so much, now that he had attained what he had supposedly wanted and wished for: a place in history and proof of the worth that he had always wanted to see in himself and had always wanted others to divine. He would set and dictate the terms.

There was less hesitancy in the words of those around him, especially those of his father, who beamed with delight and brimmed with satisfaction. Shortly before his son's inauguration, former President Bush could not get through his planned remarks at a charity dinner in Houston because he broke down crying. At a subsequent event in Washington, a private black-tie reception in his honor, he gave a similarly sentimental speech.

It began with a nod to the event's host, James A. Baker 3d, a veteran of his administration who had led his son's legal team during the Florida recount. "You know," the former president said, "I screwed up a lot of things when I was president but the first thing I did I got right . . . The first thing, the night after we were elected, in Houston, was to name Jim Baker secretary of state.

"He has been at my side politically," Bush continued. "He and I have done many other things together. Shooting birds in south Texas was one of the most fun things and so many others." He noted that when Baker had introduced him minutes earlier and mentioned their many personal connections, "He left out the fact that he and I were the men's doubles champions at the Houston Country Club for two years in a row. Nobody cares anymore. They took the flag down that had our name on it. It's gone."

He remarked that people were beginning to refer to him as "Number 41," because he was the forty-first president of the United States, and to his son as "Number 43." "I used to be George Bush," he said. "I used to be President Bush. Now I don't know what the hell I am. I'm going home Sunday and leave this to Mr. Quincy." This was a new nickname he had coined for his son, a reference to John Quincy Adams.

"I have nothing else to say," former President Bush continued, "except I wrote my doctor at the Mayo Clinic and—e-mailed, e-mailed him. And I said, 'John, can you prescribe anything to dry up tears?' " He was talking at this point about what he would be doing in the morning: standing beside his son on the inaugural stage. And he added, "I hope if tears flow—the Bushes cry when they're happy and cry when they're sad—I hope people will understand that they are looking at the proudest father in the USA."

When his son left Texas to travel east toward his inauguration, he took a detour to Midland for a farewell rally, a bit of image-burnishing on his way to the White House. Once again, he was claiming that he belonged not to the corridors of power in Washington but to the hardscrabble patches of Midland. The city had printed a special commemorative magazine for the occasion, and thousands of local people stood in near-freezing temperatures, on lines as long as five city blocks, to get near the outdoor stage and a glimpse of the president-to-be. As he walked toward the microphone on the stage, he briefly donned a white cowboy hat over his business suit.

"I'm going to take a lot of Midland and a lot of Texas with me up there," he told the crowd, which roared its approval. Toward the end of his remarks—brief as always, lasting less than fifteen minutes—he pledged that he and Laura "will never quite settle in Washington," a place he had described during the campaign as a partisan snake pit, a place he had said he could take or leave. And then, when he finished, the first blast of music that came from the enormous speakers near the stage was the refrain of a Ricky Martin song, and the first words in the lyrics were these: "Do you really want it?"

CHAPTER TEN

PRESIDENT BUSH PLOTTED HIS RETURN TO Crawford within a month of his inauguration. He had worked it out perfectly. On a Friday morning, he would leave Washington on his first foreign trip as president and head to Mexico, tracing a route that almost passed right over his beloved patch of Texas. Around midday Monday, he would attend an event in Oklahoma City, much closer to his ranch house than to the White House. The itinerary nearly cried out for a weekend layover in his Texas home, and he was ensconced there by nightfall on Friday, having limited his adventures south of the border to seven hours. He spent the next sixty hours in blessed rural repose.

Laura Bush spent two of her first four weeks as first lady there. She had her reasons: the inauguration had come before the couple was able to install and arrange all the furniture in their new house on the property; there were kitchen cabinets to be filled, drawers to be stocked, door knobs to be tested, knickknacks to be arranged. Still, her absence was odd enough, so soon into her husband's first days in office, that a Democratic rumor mill went into overdrive, planting questions about whether there was marital strife, whether a spat had sent her packing. It hadn't. She simply had interests well beyond the Potomac, and she wasn't going to be self-conscious about indulging them. She had parameters. Might as well establish them from the start.

Her husband followed suit, starting with the hours just after his inauguration. Bowing to custom, he set out for a round of inaugural balls and danced with Laura at each, but not for long. The Bushes clocked 29 seconds of dancing at the first ball they attended, 46 seconds at the second, 56 seconds at the third, 67 seconds at fourth, 50 seconds at the fifth, 48 seconds at the sixth, 38 seconds at the seventh, 50 seconds at the eighth and 49 seconds at the ninth. They were back in the White House at 11:37 P.M., which was more than an hour past Bush's usual bedtime but also more than an hour ahead of the schedule that had been outlined for him. Thus commenced what was initially shaping up to be the punctual presidency. The perfunctory presidency. The parsimonious presidency, at least when it came to the duration of Bush's public appearances. Many of his speeches lasted less than fifteen minutes.

His interest in spending a working weekend in the White House was demonstrably minimal. When he wasn't in Crawford, where he fled about every six to seven weeks, he was usually at Camp David, where he sometimes arrived as early as two or three on a Friday afternoon. To foster good relations with his cabinet, he often invited one of his cabinet secretaries and his or her spouse up to the Maryland retreat, but the purpose wasn't policy discussions. He gave his guests maps of the property and sent them on their bucolic way, so he could spend ample time relaxing on his own. Both there and in Washington, he held fast to his fitness routine, often interrupting his day for it. Shortly before noon one day in February, a gunman outside the White House fired off a few shots, and the press later demanded to know where the president and vice president had been at the time. Cheney was in the West Wing—working. Bush was in the private residence—working out.

Bush's larger responsibilities did not erase smaller concerns. A few Democratic aides emerged from a meeting with Bush in the White House on his first full workday as president shaking their heads. The subject of the Florida vote count and the need to pursue electoral reform had come up, and Bush had assented that the issue should be addressed, then segued into a more personal observation. Those reckless newscasts, the ones that had prematurely taken Florida away from him—something had to be done about *that,* too. They had spoiled his Election Night merriment. They had ruined a nice family dinner. He seemed fixated on the Bushes' aborted feast and media-induced indigestion. It was wrong, he told the congressional leaders gathered around him. It was just plain wrong.

While the heavy hitters whom he had installed all around him—Cheney, Treasury Secretary Paul O'Neill, Chief of Staff Andy Card, Defense Secretary Donald Rumsfeld, Secretary of State Colin Powell, National Security Adviser Condoleezza Rice—plotted big initiatives and thought big thoughts, Bush floated both above and beneath it all, at once helium-buoyant and resolutely down-to-earth. He invited members of the Kennedy family to the White House for a screening of *Thirteen Days,* which revisited former President John F. Kennedy's role in the Cuban Missile Crisis, and served them hamburgers and hot dogs. He invited Representative Dick Gephardt, the leader of the Democratic minority in the House, for a lunch on Gephardt's birthday, served him a cake and gave him a copy of the DiMaggio biography as a gift, signed from one baseball fan to another. He fretted over the transition that Spot and Barney were making into life in the White House. When a reporter asked him how his family was adjusting, Bush first talked about Spot and how well he seemed to be getting along. Only a few sentences into his response did Bush mention Laura. She was fine, too.

Meetings with foreign leaders might well have been occasions for his most dignified self, but he laced them with a levity that was unusual. During the trip to Mexico, which placed him at a news conference with the country's leader, Vicente Fox, Bush played his peculiar brand of footsie with reporters as the television cameras rolled. He referred to Steve Holland, the lead White House correspondent for Reuters, as "Estefan," and he indulged his nickname for a giraffe-tall reporter for the Bloomberg news service, calling on "Stretch." At one point, when it was Fox's turn to speak and he was in the middle of responding to a Mexican journalist, Bush gave me a smile and a nod and silently mouthed a question: Was I itching for an opportunity to ask him something? I shook my head no, perhaps a bit too subtly, and so he repeated his offer, but this time with unintended sound effects. As Fox continued to speak, the microphone under Bush's chin picked up his lowered but distinctly audible voice. "You want to go next?" he said.

A month or so after that, when Bush was meeting for the first time with the German chancellor Gerhard Schroeder, at the White House, he introduced Schroeder to the White House press secretary, Ari Fleischer, by commenting that Fleischer had "a face only his mother could love." This happened off-camera and was not necessarily meant to be overheard, but that was not the case with Bush's jesting the following week, during a joint appearance at the White House with Egyptian president Hosni Mubarak. The context of both the meeting with Mubarak and other events that day was grave indeed. Violence in the Middle East raged on, and an American spy plane over international waters had collided with a Chinese fighter jet and been forced to land, leaving twenty-four American service people in the hands of the Chinese military. The reporters who were given a few minutes with Bush and Mubarak wanted comments and answers from Bush, and as he tried to cut through the overlapping voices, he asked one journalist: "Did Gregory steal your question?" It was a reference to David Gregory, who covered the White House for NBC, and Bush then turned to Gregory and said, "That's very inhospitable of you." The White House transcript of the event noted, in parentheses, that there was "laughter" all around.

As the months went by, there were more jokes, and Bush deliberately made himself the butt of many of them. He went to the annual dinner of the Gridiron Club, a Washington press society, and kidded about his joy over the mapping of the genome, which he said raised the possibility of cloning Cheney and not having to do any work at all. He went to the White House Television and Radio Correspondents dinner and read aloud from a book that had collected his most absurd gaffes, malapropisms and other bits of grammatical and syntactical nonsense. During a speech to graduating students at Yale University, he humorously recalled his own days there as a beer-guzzling rapscallion, not a book-devouring Renaissance man, making no bones about it.

"To those of you who received honors, awards and distinctions," he told the young adults all around him, "I say, 'Well done.' And to the C students, I say, 'You, too, can be president of the United States.' " He went on to state that in later years, when they might make their own visits back to Yale, "If you're like me, you won't remember everything you did here." One thing Bush did remember was that he and one of his classmates, Dick Broadhead, now the university's dean, "put a lot of time in at the Sterling Library. We had a mutual understanding—Dick wouldn't read aloud, and I wouldn't snore."

It was undeniably funny. It was also startling, in that it served as a reminder of the odd way Bush fit into his new role. But it was something else, too: shrewd. One of Bush's political talents was his ability and willingness to recognize what kind of persona he could successfully fashion and what wouldn't work, what he could tenably claim to be and what he could not. Another was his way of neutralizing or at least weakening his critics by beating them to the punch, and his self-deprecating comedy routines accomplished that while they also showcased and celebrated one of his charms: his comfort with—and lack of serious insecurity about—who he was.

Through it all, what Bush frequently seemed to be doing before September 11, 2001, was making the presidency, in a sense, smaller—because it suited him better that way, because he wanted to, because it was the inevitable consequence of his character. He was stripping it of some of its pomp and diluting some of its demands, so that it was less stuffy and more manageable. He was collapsing it to his own preferred contours, recreating the office—as had every president before him—in his own image, according to his strengths and weaknesses, likes and dislikes. He was finding the right place and profile for his own personality in a new milieu, an adaptive alchemy he had honed over a lifetime.

For several weeks into his presidency, whenever he entered a room, he did not have a band play "Hail to the Chief," an homage that was like oxygen to President Clinton. And while this humbler posture partly reflected the election's messy end and a rise to power that did not warrant crowing, it also reflected Bush's lack of affinity for ceremonial trappings, his advisers' understanding that those trappings sometimes dwarfed him and his desire to give the presidency a less stilted aura. So did his decision to hold his first White House news conferences not amid the moldings and chandeliers of the august East Room, where such events were traditionally held, but amid the fluorescent lights and general chaos of the press room, a shabbier, plainer setting. Unlike Clinton, he did not try to grab headlines and the top of television newscasts if they did not naturally come his way; he did not feel that his perspective and voice needed to be added to any and every unfolding public drama. Within a span of two weeks in March of 2001, there were shootings in two schools in Southern California, and the sum total of Bush's reaction to the first one was a few sentences in response to a journalist's question, while he publicly said nothing at all about the second one. Clinton would have had an empathetic exegesis at the ready.

Photographers complained that at some of his minor appearances, it was harder than it had ever been with Clinton or with the first President Bush to get a good picture, because Bush flitted to and away from the microphone so quickly and did not insist on owning the foreground. Audiences in a few instances flashed expressions of surprise and disappointment when Bush took his turn on the dais at an official Washington social event, because the new president turned in performances that seemed deliberately tepid. At the annual President's Dinner at the Washington Convention Center in late June, he told thousands of guests that he and Laura would not be staying for the meal, because "we've got to go home and feed Barney the dog."

That is not to say he approached the presidency lightly or with any disrespect. Quite the opposite. He was the grandson and son of proud politicians, and while that ancestry had perhaps demystified the presidency to the point where he, unlike most people, could imagine attaining it, that ancestry had also filled him with a profound, personalized patriotism that coursed through much of what he did. Even after many months in the office, he could not stop pinching himself, misting up and reminding himself that he was in possession of something delicate and sacred: the people's trust.

Senator Olympia Snowe, a Maine Republican, told me how impressed she was by a conversation that she had with Bush aboard Air Force One nearly two months after his inauguration. He was still talking about the magical feeling of taking the oath of office, she said, and he remarked that he had been so wrapped up in that heady moment that he had not even noticed what a miserably gray, drizzly day it was. Another two months later, Bush told a group of business people who had gathered for a special dinner in Washington: "It's hard to describe the honor I feel every morning, walking into the Oval Office. I'm confident that my last day in office will be like my first. That Oval Office just inspires, I think, an awesome sense of responsibility."

He told friends even more than that. He recounted for them a moment during his private meeting with dozens of foreign leaders at an international conference in Quebec in April, when the president of Chile asked him to speak a bit longer than he had. Bush demurred, saying he had talked long enough and that the other heads of state should take their turns. "But," the president of Chile responded, "you're the most powerful man in the world." And Bush realized anew that he was, and it hit him with a heart-stopping force. He also shared with friends his experience, earlier, when he walked among the troops at Ft. Stewart in Georgia, locked eyes with several of them and saw something gripping there. "I think they were sending me a message: 'If you're going to send us to war, you better be damned sure you know what you're doing,' " Bush recounted. This, too, filled him with awe, at a point when he had no reason to believe—and no way to foresee—that he would be confronted with precisely such a decision.

What had always been true about him remained, in his presidency, perfectly preserved. He was at once enraptured with his station in life and determined to bring some insouciance to it, to hold onto his inner imp. One of his aides told me about the time, right at the beginning of his administration, when Bush peeked into the early-morning meeting of his senior staff, where the eighteen to twenty most influential White House officials gathered to discuss the day's business. They were sitting and chatting when the door abruptly opened and Bush stuck his head through it. They immediately stood in deference to the presence of the president.

"Sit down," Bush told them, shrugging off the courtesy, and then explained, in a kidding manner, that he simply wanted to make sure they were all working. He left.

Moments later, the door swung open and he was back.

His aides stood anew.

He smiled, chuckled and told them that he just wanted to see them do that again—that it gave him a charge.

Many weeks after that, Laura Bush arrived at the Oval Office with several acquaintances to introduce them to her husband. The appointment was on his schedule, but his previous meeting in the Oval Office was running late, and Laura waited at the door for several minutes. Logan Walters, Bush's personal assistant or "body man," did what he usually did: went briefly into the office to tell the president that his next visitor had arrived, thus giving Bush a way to cut off and shoo away the people who were already there.

But it didn't work, and so Walters, in concert with Laura,

found Barney the Scottish terrier, affixed a note to him that said that the first lady was still waiting, opened the door to the office just long enough to drop Barney inside and then quickly shut it. They waited to hear Bush laugh, which he inevitably did. He could be depended on to find the little stunt hilarious.

As it turned out, Bush was having a good time. While the pursuit of the presidency had been taxing and the prospect of it hadn't lit inside him the kind of fire that raged in more nakedly, drivingly ambitious men like Clinton and Gore, the actual reality of the presidency could be awfully pleasant. He had no higher authority, save God; could delegate plenty; and had vastly more control over his schedule than he had ever had during the campaign. Travel could often be confined to short day trips. He could assume the role of Baseball-Fan-in-Chief if he wanted to, which of course he did. When spring arrived and the boys of summer came out swinging, he went to Opening Day at a new stadium in Milwaukee, then invited a big group of Baseball Hall of Fame members to the White House for a special lunch. He announced that the South Lawn would become a sometime site for Sunday T-ball games for Washington children. At the first game, he spoke for less than five minutes and then tucked himself inconspicuously into the bleachers and simply enjoyed being a spectator. The San Diego Chicken, a man in poultry garb, was on hand, and Bush exhibited a special delight when this famous baseball mascot shoved a ball down the throat of his costume, and, seconds later, let two fall from his rear.

Months into the presidency, he was still raving about all the cool extravagances and gadgets, like the made-to-order food from the White House kitchen and the little red button in the private dining room off the Oval Office that he could use to summon the butler. There was an outdoor track and, near the Oval Office, a shrubbery-enclosed garden that Bush decided was the perfect place to cool off—or "sweat down," in his words—after a run. Sometimes, in a multitasking vein, he would use the minutes during which he sweat down in his damp, ripe workout gear to meet with a top aide, recovering from running and running the country all at the same time.

And then there was Air Force One, a behemoth of a 747 outfitted for his comfort and enjoyment. In the second week of June, he granted me an interview in his spacious office toward the front of the plane, and I stepped inside to find him still wearing the jeans and sport shirt he had donned for an outdoor event just hours earlier in the humid Tampa heat. I made a comment about the nice digs, and he beamed, like a little kid with a nifty toy. He told me that I should go take a look at the rest of the quarters. I declined, saying I knew he was eager to be done with the interview and didn't want to hold him up. But he would not be dissuaded. He shot up from the chair behind his desk and insisted on giving me the tour himself, leading me to the nose of the plane, where a tapered room included a double bed that could be used on long flights. Then he pointed out his private bathroom, complete with its own shower. His tone of voice seemed to say: Is this cool or what?

He would not be brought down from that high. Once the formal part of the interview began, I asked him if there were times, like the days when the spy plane and its crew were being detained in China, that put a knot in his stomach.

He shook his head. "A president is—there's a lot on my plate on a daily basis," he said.

"That would suggest the knot would be constant," I responded.

"Well, there are some things over which I have no control and some things I can influence, and I'm able to distinguish between the two." This sounded like a variation on a saying that might be embroidered, framed and hung above a fireplace, and I chalked it up then as a platitude, a cluster of easy words to fill up the space and run out the clock. Only later, when I watched him in the aftermath of the terrorist attacks, did I also see something true and meaningful in his description of his outlook on the presidency and on life, an explanation for his ability to ride out storms that might lay waste to someone with a less keen sense of destiny and a less ready acceptance of fate. Bush's serenity wasn't that of the ignorant or aloof. It was that of the faithful, and as subsequent events would show, it didn't mean that Bush wasn't prepared, when it became necessary, to fight and to try.

Just a few weeks before the interview, Senator James Jeffords, a Vermont Republican, had left the party and become an Independent aligned with Democrats, giving that party control of the Senate. Jeffords had explained his decision by saying that Republicans—and the Bush administration—had become too conservative. I asked Bush about polls that suggested that a plurality of Americans, including some Republicans, felt that Jeffords was right. But when I asked Bush questions about all of this, he refused to look backward, dismissing any disappointment, concern or sense of responsibility for what had happened. He wanted to look forward, and what he insisted on seeing was his continued ability to herd legislation through a much rockier political landscape.

Surely, I pressed, the day that Jeffords left the Republican party must have been "one of your bluer days in office."

"I haven't had any blue days," he said, with an utterly straight face. "Every day has been chipper."

"That," I asked, "was a chipper day?"

"Every day is a great day when you're the president," he said, and if there was any sarcasm in his voice, it was subtle.

He confessed only one qualm. "It's hard," he said, "to run out of the White House and jog the Mall." He had never done that.

This was the all's-right-with-the-world attitude he took with him a week later when he went on his first overseas trip as president, a visit to Europe that would take him to five countries in five days. It was his longest, boldest stride onto the world stage, and it was met with wariness and reservations by his immediate audiences. He left Washington for Madrid on the day of the execution of Timothy McVeigh, a widely covered story that rehashed Bush's record in Texas as a fervent advocate of capital punishment, a practice no longer used by European governments. And he arrived the next morning on a continent deeply suspicious of his conservative politics and many of the policies he was pursuing. He had abandoned the Kyoto accord on global warming, which many European governments at least publicly supported. He had made clear that he wanted to proceed with the construction of a missile defense shield, even though two of the United States' key European allies, France and Germany, were protesting that strategy.

He had also agreed to meet Russian president Vladimir Putin in Slovenia, a country he had once, during the presidential campaign, confused with Slovakia, a very different beast. That old gaffe was of course finding its way into much of the advance coverage of his trip in the news media, and it was bound to do so at several points during his trip as well. Bush had not paved an easy road for himself.

The night before he touched down in Madrid, I landed there with the rest of the press corps and found myself back in the "bubble," international-style. It was the weirdest brand of foreign travel imaginable. Upon departure, at Andrews Air Force base, you were handed a brick-thick, concrete-heavy briefing book that was like a foreign-policy primer and a Lonely Planet travel guide rolled into one daunting, chartreuse-covered tome. And you handed over your passport to a White House travel official, after which you never saw it—or were asked for it—again. You never went through customs or immigration. In each new country, buses lined up on the tarmac, a few yards from the plane, and you hopped on, went to a hotel, were given a room key and were never forced to have a strained conversation with a desk clerk or taxi driver who didn't speak English. You didn't even have to convert dollars into another currency, because most everything had already been charged—or would later be charged—to the office or a credit card, and each hotel had a filing center stacked at regular intervals with food: paella in Spain, smoked salmon in Sweden, kielbasa in Poland. It was Europe on 10,000 calories a day.

That is not to say it was easy, or even pleasant. Security was so tight that we were all usually hustled to Bush's daily speeches and news conferences as many as three and no less than two hours before the actual event. When we weren't in buses that looked exactly, right down to the kaleidoscopic upholstery, like the ones we had used on the campaign, we were in filing centers that might as well have been in Des Moines or Detroit. We couldn't stray far from them, because this was where administration officials might or might not drop by to "brief" us, a terribly useful and convincing exercise in which they told us how much European leaders were coming to adore Bush. Our handlers—many of the same fresh-faced twentysomethings from the campaign—herded us from one European city to the next early in the morning, which meant that baggage call was usually at 4:30 or maybe 5:30 AM. But we usually weren't through with our editors on the other side of the Atlantic until one A.M., because it was six hours earlier there. This meant that we got about three hours of sleep each night, less if we treated ourselves to late-night cocktails, which a number of us did in Sweden, at the end of the third day. It crunched our time in bed to ninety minutes, and by the fifth day of Bush's trip, when he met Putin, just about everyone in the press corps was walking around with the vigor, animation and mental acuity of mummies.

It was an odd experience all the way around. In Madrid, the chief press aide for Condoleezza Rice bent down beside me in the filing center and whispered an invitation to join a "small group" of reporters for an off-the-record dinner with Rice the following evening in Brussels. Midway through the next day, I had heard nothing more and asked the aide, Mary Ellen Countryman, whether, where and when this was happening. She said I would receive "an envelope." It was dropped on my laptop a short while later, and it was furtive in the extreme: no names, no outright acknowledgment of the nature of the meeting, just a printed advertisement for a nearby restaurant and, in tiny handwriting, the notation "8:30." I showed up at the place and time in question to find Rice, Hughes, four other journalists and several empty seats. One belonged to a magazine writer who had not been told how she would be informed of the details of the dinner, found the strange envelope under her hotel-room door, did not see the "8:30" and figured the picture of the restaurant was some kind of solicitation. She threw the paper away and made her own plans for the evening.

It was undoubtedly an odd experience for Bush, too, although he really didn't show it. You could take the president out of his usual surroundings, but you couldn't take the unusual personality out of the president. Whether he was "doing well" on his Europe trip was a topic that journalists chewed over endlessly, and rather pointlessly: we weren't privy to his most substantive conversations with foreign leaders and the fruits of those discussions wouldn't be apparent for some time to come. But what was immediately evident, in the public glimpses of Bush that we got and some of the details I gleaned from his aides, was that his colloquial, goofy, verbally clumsy style had survived the transatlantic flight intact.

It was in full flower just before he left, when he sat down a few hours before his departure with a group of journalists representing European publications. A transcript was provided to us the next day, and it was rife with Bush's peculiar locutions, like his description of Cold War détente and the thinking behind the Anti-Ballistic Missile Treaty as the doctrine of not "blowing each other up" and his musings on one of his destinations. "I'm the first president, evidently, to have gone to Sweden," he said. "I look forward to going to Sweden. My state of Texas has got a lot of Swedish-Americans: Swedish-Texans." Demographers, take note. There's a new census category in town.

In that same interview, he jumbled the middle consonants in the last name of the Spanish prime minister Jose Maria Aznar, calling him "Anzar." Bush had made the appropriate correction by the time he stood with Aznar at a news conference on the next day, Tuesday, in Madrid, and he had decided to relate to his newly met acquaintance as a bosom buddy. "Good job," he told Aznar at the end of their joint conference, the way a football coach might praise a quarterback who had just completed a four-yard screen on third-and-one. When Bush ran into British prime minister Tony Blair at a NATO meeting in Brussels on Wednesday, he called out, "Hello, Landslide!" It was a towel-snapping reference to Blair's recent reelection victory and a greeting much more casual than the ones Blair was accustomed to receiving.

At a subsequent press conference with Lord Robertson, the secretary-general of NATO, Bush mistakenly referred to him at one point as "Robinson" and came out with an ear-catching redundancy when he stated his belief that he had shown European leaders the wisdom of missile defense. He said they now understood "the logic behind the rationale."

On Thursday, in Sweden, he opined that Europe needed "more countries," an ambiguously worded wish that probably had to do with the growth of the European Union but came out sounding like a recipe for carving the continent into Andorra-sized McNuggets. Striving for a note of sympathy about the AIDS pandemic, he also said, "Africa is a nation of incredible disease." On Friday, during a speech in Poland, he was supposed to say: "Today, Poland's own Golec Orchestra." He instead said: "Today's own, Poland's orchestra, called Golec's."

Just as he had done down in Mexico, with Fox, he maintained his habit of flirting with reporters in front of open microphones and live cameras, infusing otherwise formal moments with playful admonitions about unwelcome questions or teasing vows to cut unruly reporters out of the loop. It was as if he had forgotten that this would be the grist for many Europeans' first impressions of him—that what they had doubted was his capacity to be thoughtful and what he was showing them was his inclination to kid around. In the middle of a press conference in Warsaw with Polish president Aleksander Kwasniewski, Bush interrupted a question by NBC correspondent David Gregory to compliment him on his threads.

"Your outfit is mighty sharp today," Bush said. He then turned to Kwasniewski and added, "Don't you think?"

Bush was on more straitlaced behavior the next day, at his joint news conference with Putin, but he nonetheless exhibited an arguably callow way of thinking and talking about things. Searching for a way to capture his stated feeling of confidence about his and Putin's future ability to work honestly together, Bush said, "I looked the man in the eye. I was able to get a sense of his soul." And there were similarly gee-whiz accents to his remarks, when he got back to Washington, to the Republican speechwriter Peggy Noonan in an interview in the Oval Office.

Noonan wrote about her conversation with Bush in the *Wall Street Journal,* noting that as Bush recounted his meeting with Putin for her, he remembered that Putin at one point said, "I love history."

Bush told Putin: "You know, it's interesting. I do, too. I like history a lot."

"You know, sometimes, when you study history, you get stuck in the past," Bush recalled saying to Putin. "President Putin, you and I have a chance to make history. The reason one should love history is to determine how to make good history. And this meeting could be the beginning of making some fabulous history."

"It's negative to think about blowing each other up," Bush later continued. "That's a Cold War thought."

Like so much of what Bush said, there was nothing wrong with the content of it, and in other stretches of the Noonan interview, he sounded rather articulate. But he had—and, seemingly, would always have—this way of expressing even some of his soundest thoughts with a strangely chatty, see-Dick-and-Jane inelegance, at least when he was speaking without notes. On his second overseas trip as president, also to Europe, in mid-July, his response to the hordes of angry protesters and the European leaders who disagreed with his policies was: "I know what I believe, and I believe what I believe is right." Less than a week later, back in the United States, he described the country's troubled economy as "bumping along." These were not quite like any presidential pronouncements Americans had ever heard before.

They seemed not to know what to make of him, but they also

seemed not to be dismissing him, at least if the polls from May through August were any evidence. While his job approval rating fluctuated, depending on the national moment and the given pollster's methods, it held pretty solid in the low-to-mid-50s to high 50s, a respectable result. He was doing better, at least by that measure, than Clinton had during his first year. Still, there were signs of potential trouble: doubts among a majority of voters about Bush's ability to handle an international crisis; qualms among independent voters that Bush was too conservative; a widespread feeling that he favored corporations and the rich and might not understand the problems of middle-class Americans.

There was at least one other looming problem, not captured as such in the polls but evident to anyone who watched Bush closely, and it was a problem that had existed from the moment he stepped onto the campaign trail. As President Bush dealt with a range of issues broader and sometimes more complicated than he had faced as the governor of Texas, it was never clear how much he really knew and, perhaps more to the point, how diligently he was trying to amass the knowledge he might need. He and his advisers were right to say that a president depended on the expertise of the advisers and cabinet officials he had gathered around him, and that choosing them well and using their counsel wisely were as important as an encyclopedic familiarity with all the facts under discussion. Presidents before Bush had proved that an excessive attention to detail and a temptation to micromanage the affairs of government did not always work. But there were bound to be situations in which he would field warring counsel, have to look to his own judgment and be served best not simply by his gut instincts and overarching political philosophy but also by information and a sturdy command of it.

This was always the rub with Bush, the great big question

mark at the heart of the man, the angle from which opponents most eagerly attacked him and proponents most strenuously defended him. The deft political adjustments he constantly made, some of his off-the-cuff retorts in casual conversations and even his own grasp of what he could and could not ace proved that he was no dummy, not by a long shot. But neither a year and a half on the campaign trail nor the beginning of his presidency had definitively demonstrated the extent of his mastery of the most challenging matters of policy and governance. And he and his aides sometimes seemed to be elaborately constructing a universe around him that deepened the riddle and forbid its solution.

They cautiously parceled out his remarks, affixing him to his talking points, lessening the possibility of surprise, hewing to that vaunted "message discipline" that had long seemed a slightly suspicious rationale. Except for major speeches, his remarks to reporters in the Oval Office or the Rose Garden were often banal and vaporous; they dissipated on contact. When he granted television interviews, it was most likely for the softest formats possible: he gave the first extended national television interview of his presidency to the sports commentator Bob Costas, who had a new show on HBO, and the subject was baseball. When he granted print interviews in connection with his first hundred days in office, he kept them to fifteen or twenty minutes and unleashed long statements that took up much of the time and diminished his exposure to unanticipated questions.

If he and his handlers loosened the tethers on him, it was often in circumstances and cases where he was bound to be most comfortable and sure-footed. And so it was impossible not to wonder: did he talk more voluminously about education because it mattered more to him than anything else, or because it was one of the few issues he had nailed? Did he make that first trip to Mexico and stress the importance of our relations with our southern neighbor because he fervently believed it, or because he spoke more than a dollop of Spanish and his years in Texas had given him a familiarity with that country that he lacked with almost any other? I remember watching him stride to the twin lecterns on Fox's ranch in rural Mexico on that afternoon in February and thinking that Bush and his aides had arranged it all so conveniently. They had found for Bush a foreign leader who had also spent more of his life in the private sector than in public service. They had put Bush together with another politician who liked to flee the capital for the parched earth of a remote hideaway and who was as fond as Bush was of cowboy boots and casual attire—both men stripped off their jackets and neckties soon after they met. They had set it up so that Bush, on his first trip as president outside the country, came face to face with his own mirror image.

I also remember that when the talk at that news conference turned to Iraq and to the bombs that American pilots had just dropped to enforce the no-fly zone, Bush turned tentative, his voice less sunny, his words more labored, his effort to flip through the index cards in his mind more obvious. This was often the case when Bush was yanked away from his most practiced script, as he was at the Democratic retreat in Pennsylvania. At that meeting, Representative Nancy Pelosi, a California Democrat, grilled him about his decision, on his first full workday in office, to reinstate a ban on federal aid for international groups that provided abortion counseling with their own money. Bush's explanation at the time was disingenuous: He had said he did not want taxpayer dollars promoting abortion. In fact, federal law was already preventing these organizations from actually providing abortions; what Bush was trying to do was make sure that the organizations, if they continued to take federal money, didn't help women find ways to get abortions from other groups. And he was jeopardizing additional work they did to prevent pregnancy and slow the spread of AIDS.

Pelosi asked him why he would worry that such funds, used for family planning and AIDS education, would be "fungible" and end up paying for abortion-related services when he was, on the other hand, so insistent that the federal money he wanted to send to religious groups would go only to social programs, not to proselytizing. He flailed for the right response and could not immediately find it. He seemed confused. To many of the Democrats watching and listening to him, it was not at all clear that he understood the fine points of Pelosi's argument.

Then again, maybe he simply didn't understand the executive order he had signed. During a meeting with Roman Catholic charity leaders that was inadvertently broadcast in the White House press room, Bush's characterization of the order was so vague and off-kilter it was almost wiggy. The ban has long been known as the Mexico City ruling because the Reagan administration first instituted it at a conference there in the mid-1980s. As Bush talked about the issue with the Catholic leaders, he referred to "the money from Mexico, you know, that thing, the executive order I signed about Mexico City." There was no money coming from Mexico. The order had nothing, really, to do with Mexico City. The most accurate part of Bush's description was the phrase "that thing," which certainly didn't suggest that he took the matter as seriously as those who were going to be affected by it did.

At Bush's first full-fledged news conference in the White House, he trotted out all the familiar sound bites and vague generalities. He smiled a lot, confirming that while President Clinton's signature expression had been the quarter-bitten lower lip, an intentional signal of empathy, President Bush's was the mischievous half-smile, an equally intentional request for mirth. But still he stumbled, or at least wobbled, refusing to delve very deep into details. An FBI agent had just been arrested on charges of spying for the Russians, and Bush was asked if agency officials should henceforth be required to take regular polygraph tests. Bush deferred to the eventual judgments of the agency leaders who were reviewing their procedures, then threw in his own, extraordinarily provocative, inexpressibly meaningful two cents.

"We ought to be concerned about espionage in America," Bush said.

He also seemed to evade a question from a British journalist about what Bush might say to British prime minister Tony Blair regarding the European Union's proposal for its own military force, which some U.S. officials saw as a potential threat to the NATO alliance. Bush said he was excited about meeting Blair at Camp David that afternoon, that they would talk about NATO, that they would share a meal. He then offered to answer the British journalist's question later in the day, after his encounter with Blair. While administration officials said the reason for the stalling was Bush's reluctance to telegraph his intentions, it seemed just as likely that he had not been prepped on a response.

That incident and Bush's fumbling in describing his executive order led me to ask several administration officials how reporters and, more to the point, voters were supposed to know and be reassured that Bush fully understood the issues and the pieces of legislation with which he was dealing. The superficiality and ambiguity of his responses left room for doubt. The answers I got were frustrating and basically boiled down to the same assertion that Bush's advisers had made during the campaign: that discipline—not confusion—was the reason for any reticence and that Bush was being admirably cautious, not alarmingly clueless.

As for Bush's terseness and desire to speak more generally than specifically, they said this reflected a praiseworthy and effective executive style that focused on the forest, not the trees, and did not mean that Bush was really lost in the woods. Vance McMahan, who had helped develop policy in the Texas governor's office and went to work under Rove in the White House, called me with a personal testimonial. "There was a meeting in the State Capitol one time in the middle of a drought," McMahan said, adding that the engineers and wonks "flashed a bunch of graphs and charts and they were talking about whether the state should build this reservoir. After that went on for fifteen minutes, the governor said, 'Tell me how much water we have, how much water we're using and how much water we need.'" This was, supposedly, an example of Bush dynamically cutting to the chase. But was it really an example of impatience with even the kinds of details that might matter? McMahan told me that he had seen Bush participate "in policy meetings that I expected to last twenty minutes but that lasted ten minutes." This was, supposedly, an example of Bush's efficiency. But I found myself dwelling on something other than his ability to cut the expected meeting time in half.

It was only going to consume twenty minutes in the first place?

CHAPTER ELEVEN

ON SEPTEMBER 13, THE TWO SENATORS FROM New York, Chuck Schumer and Hillary Clinton, and the two senators from Virginia, John Warner and George Allen, visited President Bush in the Oval Office. Their states, where the World Trade Center and the Pentagon were located, had been the most directly and immediately affected by the terrorist attacks, and the senators were there to discuss federal aid. They wanted as much as possible. They were counting on just a few minutes of the president's increasingly precious time. But they got not only a commitment, which later turned out to be more complicated and uncertain than it seemed at first, of $20 billion. They also, to their surprise, got a half hour with Bush, who seemed in no hurry to dispatch this obligation, one of umpteen entries on the day's lengthy to-do list.

Before the terrorist attacks, Bush had seldom attended the "message meetings" at which the administration's communications strategy was considered and plotted. But soon after, he participated in one at the start of each day, joining heads with Hughes, Rice and others to attend to the important business—not just for him, but for the country—of setting the right public tone and expectations. It was Bush, from the beginning, who felt most forcefully that Americans needed to be conditioned into thinking about the coming war against terrorism in patient terms, and that unequivocally correct belief informed almost every public statement he made, even the ones that were delivered more or less spontaneously. You could not listen to him over time and fail to suspect that he understood, regardless of how prudent or effective a response he mustered, the enormity of what had befallen the country. Or that he understood how desperately Americans were looking to him for leadership.

A long telephone conversation that I had with someone close to him—someone outside the administration—strengthened this impression. This person spoke in a hushed, grave voice as he talked about what he could see and what he knew to be happening to Bush over the days following September 11, a period during which, he said, the full force of a president's responsibilities was hitting Bush like never before. He said that Bush realized that everything in his life to this point paled in importance next to the current moment and challenge and that Bush had come to believe that his actions from September 11 on would define not only his presidency but, really, his time on earth. He said that Bush felt he had "begun a new life that is inextricably bound to September 11 and all that it implies." Mulling over those implications in a way that clearly reflected Bush's own thinking, he said: "There's no question what Bush's legacy will or won't be. He either beats this back—or we lose."

Even Schumer, an often partisan Democrat who had led the unsuccessful Senate fight against the confirmation of John

Ashcroft as attorney general, told me that he was seeing new, deeper shades in Bush. There were undoubtedly some self-serving political calculations in this testimonial, because Schumer and Clinton were both trying to mix honey with vinegar in their appeal to secure as much federal succor for New York as they could. But Schumer's words seemed to go beyond that, and it would have been risky for him to fabricate bits of conversations he had had with the president. What he remembered the president saying to him about the aftermath of September 11 was that his life's work would never be the same. "He has told me several times that he is staking his entire presidency on this—that the mark of whether he's successful is whether he can succeed in his goal of wiping out terrorism," Schumer recalled.

Schumer and other lawmakers who had spent time with the president since the attacks said that he was more attuned to details, more eager for information. Aides conceded that Bush still wasn't given to meetings that went on too long, but in the new context of rapidly unfurling events, that tendency was understandable and possibly useful: there were innumerable aspects of the government's response that needed consideration, and none of them could be given exclusive attention. Aides said that he was asking more questions in those meetings, grilling his advisers with more requests for explanations and often demanding to talk not only to the deputy from an administration agency who was giving him a briefing but to the head of the agency. They said that he was calling world leaders early in the morning and late at night and had spent forty minutes on the phone with Putin at one point. Of course there was aggressive spin at work. But the thing about spin was that it had to be tethered at least somewhat to the truth or, somewhere down the line, it might be revealed as utterly mendacious and backfire. And if the truth with which aides were working was not all that compelling, the spin might well end up looking silly in the extreme.

The events surrounding the grounding of the China spy plane back in April of 2001 had proved as much. In the beginning, Bush said very little. Twice in two days, he appeared before reporters in a picturesque setting to read a statement of a few paragraphs, then turned quickly on his heel and walked away before anyone could ask him anything, before anyone could scratch the surface. Questions arose: Was he really on top of this? Was he really in charge? To satisfy reporters that he was, his advisers subsequently mined Bush's daily activities for examples of Bush's intense scrutiny of the situation, of the valuable input he was providing.

This is what they came up with: Bush had asked the American military officials monitoring the conditions of the detained service people if the captives had access to Bibles. He had asked if they were staying in the equivalent of officers' quarters and if they were able to get any exercise. His advisers also noted that while Bush had traveled to Delaware that week to speak at a Boys & Girls Club, the situation with China never left his mind, and as soon as his helicopter deposited him at the White House, he barked to his staff: "Get me Condi!" This would be Rice, with whom he talked frequently, and his interest in summoning her was supposed to be synonymous with his control over events.

Aides had more luck selling Bush's investment in what, before September 11, had actually seemed like a signal act of his presidency, his decision about whether to permit federal support of research involving embryonic stem cells. Such research held some promise for developing treatments or even cures for a range of diseases including juvenile diabetes, Parkinson's and Alzheimer's, but many social and religious conservatives who opposed abortion felt that the destruction of embryos, even those that were frozen in fertility clinics and bound to be discarded, was equally horrifying, an act tantamount to the elimination of incipient life. Bush had taken pains during his campaign and the first months of his presidency not to anger those conservatives. And yet he needed to be just as concerned about losing his tenuous grip on the political center and on his claims to a kind of compassion that defied liberal, conservative, Democratic or Republican labels. To make matters more politically complicated, many moderate Republicans and even a fair number of conservative Republicans who were against abortion favored the research and served notice in the news media that Bush would disappoint them if he forbid federal support. Advisers considered it a lose-lose proposition, with one exception: If Bush could prove that he reached whatever conclusion he did after painstaking consideration, he would at least project the kind of seriousness he often failed to.

Bush ultimately went with a conservative approach when he announced his decision in early August. While he presented it as a kind of compromise, it was not much of one: he said that he would permit federal support only for research on embryonic stem cell "lines," or colonies, that had already been derived from destroyed embryos, but not for the destruction of any more. This did not keep perfect faith with the most ardent religious conservatives, who wanted a complete ban on federal funding for the research, but it did not thumb its nose at them, either. And it left many scientists feeling cheated and shortchanged, while it left a number of moderate Republicans feeling that Bush had yet to prove to them that he had the courage or political temperament to break with the right when doing so truly threatened to cost him. "Once again, he gets a gentleman's C," one of his warier moderate Republican supporters told me. "Story of his life."

But his aides wanted to make sure he got an A for effort, and they succeeded in getting him at least a B. Hughes wrote a set of remarks, which Bush delivered on live television, in prime time, that explained his decision, and Bush used those comments as a sort of tutorial for the country on all of the arguments he had weighed, all of the science he had studied, all of the perspectives that he felt he had to respect and balance. The next day, Hughes held a briefing for reporters, during which she went through the list of high-level officials and advocates Bush had consulted, replete with names and dates and sometimes the locations and durations of the conversations in question. The goal was to present the sometimes glib, flip president as "deliberative," a word that aide after aide, time and again, used to describe Bush's approach to the matter.

It was spin at its most furious, but there were at least two reasons not to dismiss it as meaningless. One was its sheer ambition and intensity, which suggested that Bush, more than ever before, *cared* about being seen as someone at least a bit more ruminative than he usually got—or wanted to get—credit for being. So often in the past, and even in the opening acts of his presidency, he had been content to be seen as someone who shrugged off his cares and worries. He had taken a certain obvious pleasure in not bending to judgments and criteria that he deemed shallow, which was how he had deemed those first presidential debates in the Republican primaries way back in the fall of 1999. But with his decision on embryonic stem cells and the way it was unveiled, he relinquished some of his rebelliousness and defiance, and there were hints of a greater maturity in that. He acknowledged that it could be important for a president to articulate the weight of some of his burdens, not just to shoulder them on a springy frame that ambled merrily along.

The other point of interest was that the spin in this case jibed with what people outside of its immediate rotation were saying. Even after Bush had made a decision that displeased them, some of the advocates who had met with him said that he had nonetheless come across as extremely interested in what they had to say and attentive to some of the finer points of the issue. And months before he announced his ruling—and before the ink on the talking points was dry—a few of the handful of aides who could be counted on for an occasional morsel of off-message candor confided that they had never seen Bush so transfixed by a given subject.

He would bring it up in the middle of an unrelated discussion of education, asking the people around him what they thought. He would sometimes solicit the opinions of administration officials several notches below senior rank. "It's come up in economic policy meetings," one administration official told me about five weeks before Bush announced his decision. "He says, 'Hold on,' and we get back into stem cell." In the end, Bush's decision was precisely the one that another well-informed administration official had told me, a good month beforehand, seemed to be the most likely—the direction in which he had long been leaning. So it was unclear that all of those meetings and deliberations had made much of a difference, and it was possible that they were just window dressing. But it was also possible that Bush had found a new level of engagement in the work before him—that his purposefulness had changed, even if his mind had not.

After the terrorist attacks, semi-candid, unauthorized interviews with a few of the less straitjacketed White House officials ceased to exist. One aide told me that the chief of staff, Andy Card, had issued a stern directive that nobody without specific approval talk to reporters, and no White House knew how to clamp down on information and hum a chosen melody like the White House of George W. Bush. Card himself was a marvel of dutiful hero worship when it came to his statements about the president, which I had learned during a telephone interview with him in early August about the administration's record so far and its plans for the fall.

Even though Card could have, with a few well-chosen and innocuous revelations that served the administration's interests, controlled the gist of my story, he clung to generalities that any Washington veteran like him had to know would never make the headline or top paragraphs. I pressed for signals of where the administration was heading, telling him it would be great to make just a little news with the story, and he said he had already made news by pointing out that President Bush was, bill by bill, living up to every promise he had made to voters. When I told him this was unlikely to be the first sentence of my story, he said that it should, but did not expect as much from the *Times*. I knew from previous conversations with other Bush aides and with Republicans close to them that the White House wanted for starters to try to take Bush away from the lecterns at which he could look stiff or awkward and put him in more casual public settings, so I asked Card about the kinds of settings that were being envisioned. He dismissed the question by saying, in the same straightforward voice with which he had suggested the proper angle for my story, "The president is *spectacular* in every setting." Adjectives did not come much more overblown than that.

And efforts to buff a president's image and shield it from any nicks did not come more fervent than they did in the Bush administration. After several commentators raised questions about the delay in Bush's return to Washington on September 11, when he had moved around the country for nine hours after receiving word that the second tower of the World Trade Center had been hit, White House officials aggressively peddled the story that there had been an ominous, credible threat against Air Force One, compelling Bush to stay away. They scolded reporters who expressed doubts about that story; one of these correspondents, who was prominent enough to be able to depend on regular conversations with the White House press secretary, Ari Fleischer, spent several days in purgatory as Fleischer did not return the correspondent's phone calls. But neither Fleischer nor anyone else in the White House ever presented compelling evidence of the alleged danger to Air Force One and the president and ultimately dropped the claim, essentially admitting that it was incorrect. The unnecessary defensiveness and haste with which they had made it—probably few Americans outside the Beltway cared all that much about the precise timing of Bush's reentry into the White House—was the kind of thing that made an observer wonder just what kind of clothes the emperor was wearing.

White House officials latched onto two preferred adjectives to articulate the president's mindset and emotions following the attacks, and they used them to the point of absurdity. If you asked the administration's chosen spokespeople how the crisis was affecting the president, they would instantly remark on how preternaturally "focused" he seemed. Then they would say that he was, above all, "resolute," a word that stuck in the ear because it so seldom subbed for its various synonyms. If you asked a Republican who was close to the White House, and had likely been educated on the message, what struck him about Bush's behavior, the answer was that Bush seemed, well, focused. Oh—and resolute. Here was a table of poker players who had never learned about the danger of "tells," a group of amateur linguists who had never availed themselves of a thesaurus.

And yet the efficacy of even such transparent tactics soon became clear. By using this vocabulary over and over, aides lodged it so deeply in the minds of reporters that these reporters began adopting it without even realizing it. On the morning after Bush's address to Congress, stories in both the *Washington Post* and the *Times* that analyzed his demeanor used the word "resolute," without quotation marks, in the first paragraphs. The word popped up in newsmagazines, too. Homer had never given one of his characters an epithet that clung so tenaciously.

It stuck in part because the idea—although not the actual word, which was much fussier than any that Bush would naturally use—felt right. Bush did look determined, and that appearance could be trusted to some extent, because he wore his real emotions on his face. He had not looked determined on September 11, or had looked a whole bunch of other things as well: tentative, tense, shocked. But by the time he spoke at a memorial service in the National Cathedral on September 14—an astonishing event that packed his father, mother, Bill and Hillary Clinton, Al and Tipper Gore and a mind-boggling cargo of political baggage in a few tight pews—he had found something like serenity and something that looked very much like gravity, both of which served something that might well be called resoluteness. The religious atmosphere probably helped, because it was one that was familiar to Bush and gave him personal comfort, which always mattered with him. He had seldom delivered such lofty language so smoothly.

Although it did not draw much comment at the time, Bush's spirituality was perhaps crucial to his ability to handle himself so well in public—and, by all accounts, in private—after the attacks. Reporters in general tended not to look for or consider this kind of influence, because so few of them put all that much stock in religion, which they associated with extreme political views based on moral disapproval and condemnation. I confess to membership in this group. And perhaps because of this prejudice, which is probably the right thing to call it, administration officials seldom dwelled on Bush's faith when they described the way he was coping at difficult junctures. But when Bush had given up alcohol around his fortieth birthday, he had also developed a more profound belief in God, and when he talked about it, he often did so in the language of a born-again Christian. It was one of the reasons that religious conservatives had, from the start of his political career, trusted and supported him more than they ever had his father; they heard in his words and cadences certain telltale signs that reporters often overlooked. It was also an explanation for his ability during grueling passages, like the period of his campaign when Gore seemed to be pulling ahead, to surrender emotionally and mentally to fate, which he believed was governed by a wise and benevolent hand. I hadn't understood this clearly before September 11.

I did after September 11. Although many of the people close to him disagreed with the following assessment, a few said that Bush believed that the mission he was confronting must be God's intention for him, a kind of reason for being, and they said that this conviction steadied him. "I think, in his frame, this is what God has asked him to do," one of them told me. "It offers him enormous clarity." A senior administration official recalled a conversation that Bush had with a group of religious leaders who visited the White House nine days after the terrorist attacks, on the same day he would deliver his address to Congress. According to the official, one of them told Bush that his leadership at this time of crisis was part of God's plan, and Bush responded, "I accept the responsibility." His father also broached the subject of the strength his son drew from the feeling that God was closely watching. In an interview with Tom Brokaw that was broadcast on NBC, the former president said, "This thing about faith, I mean, this is real for him. This is real. Here's a man that's read the Bible through twice. And it's not to make it holier than thou or not to make a political point. It's something that is in his heart."

In fact, Bush slipped during the first days after September 11, talking about the "crusade" of the country's new fight against terrorism, a word with unsavory historical connotations that was not the right message to send to the non-Christian nations that the United States needed in its international coalition. (After input from advisers, Bush dropped the loaded term thereafter.) Bush ceaselessly referred to the terrorists and those who harbored them as "evildoers" and to Osama bin Laden as "the evil one," terms that sounded alternately archaic and apocalyptic and cast the country's struggle as one of righteousness versus immorality, the servants of virtue versus Beelzebub himself. During his address to Congress, Bush interpreted this crossroads in America's history in grand, emphatic and even Manichaean terms that suggested the influence of his faith. He said that "this is civilization's fight," that freedom and fear were at war and that "God is not neutral between them."

The address was filled with such flourishes, which Bush delivered with as much poise as he had ever demonstrated. But the prose was as striking as the poetry, casting Bush as the nation's educator-in-chief, dispensing rather than receiving lessons on geopolitics. "Americans have many questions tonight," he said. "Americans are asking, 'Who attacked our country?' The evidence we have gathered all points to a collection of loosely affiliated terrorist organizations known as Al Qaeda.

"They are some of the murderers indicted for bombing American embassies in Tanzania and Kenya, and responsible for bombing the USS *Cole*," Bush continued. "Al Qaeda is to terror what the Mafia is to crime. But its goal is not making money; its goal is remaking the world and imposing its radical beliefs on people everywhere. The terrorists practice a fringe form of Islamic extremism that has been rejected by Muslim scholars and the vast majority of Muslim clerics. A fringe movement that perverts the peaceful teaching of Islam. The terrorists' directive commands them to kill Christians and Jews, to kill all Americans and make no distinctions among military and civilians, including women and children.

"This group and its leader, a person named Osama bin Laden, are linked to many other organizations in different countries, including the Egyptian Islamic Jihad and the Islamic Movement of Uzbekistan. There are thousands of these terrorists in more than sixty countries. They are recruited from their own nations and neighborhoods and brought to camps in places like Afghanistan, where they are trained in the tactics of terror. They are sent back to their homes or sent to hide in countries around the world to plot evil and destruction. The leadership of Al Qaeda has great influence in Afghanistan and supports the Taliban regime in controlling most of that country."

Listening to him, I was struck anew by the oddity of his journey, by how unlikely it was that Bush had ended up where he was. Only two years earlier, he had still been receiving foreign policy tutorials from the various experts that his advisers had assembled to bring him up to speed. So many times during his campaign, he had tripped over or blanked on the names of foreign countries and the names of foreign leaders. He had often seemed least surefooted when it came to global issues, and had rooted his bid for the presidency in two rather basic, and not especially revolutionary, domestic ideas: a tax cut and reforms for the country's public schools, which received only seven percent of their money from the federal government and were shaped primarily by state and local authorities. And now he was putting himself forward as the one who would teach the rest of us about the world at a time when it was maddeningly difficult to understand it and urgent that we did. Bush had taken an unusual trip to what seemed, for him, the unlikeliest of destinations.

The depth of his own understanding was impossible for any outsider to gauge. But there were other aspects of his response to the crisis he was facing that could be measured. One was his patience, which turned out to be remarkable. He waited almost a full month after the attacks to send the first war planes to Afghanistan and drop the first bombs, resisting any emotional or political temptation to strike sooner. That period of time enabled him to put together a complex international coalition unlike any since World War II and to make absolutely sure that our key allies were firmly on board. Bush could also be judged on his flexibility, which turned out to be considerable. In assembling the coalition, he abandoned some of the unilateralist rhetoric and tendencies that had informed his push for a national missile defense shield and his resistance to an international pact to fight global warming. He demonstrated a new appreciation for the interlocked fates of free nations.

Perhaps most strikingly, he began, in terms of the future of Afghanistan, to endorse the kind of "nation building" that he had so harshly disparaged before September 11. Bush had long said that what was wrong with American foreign policy during the Clinton administration was the way American forces in Somalia and the Balkans had entered into such a mission, a prolonged commitment that he said had nothing to do with defending the United States. The second of his three presidential debates had turned into a fierce disagreement on the matter with Vice President Al Gore, who believed that nation building could be vital to preventing future wars, and Bush refused to give any ground. Two weeks after the terrorist attacks, Bush still wasn't budging. He said emphatically, "We're not into nation building. We're focused on justice."

But one month after the attacks, during an evening news conference at the White House, Bush signaled a change of heart, saying that the United Nations, with the participation of the United States, should "take over the so-called nation building—I would call it the stabilization of a future government" in Afghanistan once the military strikes were done. He had recognized that without an eventual commitment to make sure that a post-Taliban government did not disintegrate quickly, there was no guarantee that the wretchedly impoverished, fractious country would not become once again what it was now: a refuge and cradle for extremists and terrorists. He had learned something about the immediacy and complexity of America's stake in distant lands, and he was willing to bend to it.

Bush had bent before during his presidency, dropping a demand for vouchers in legislation for education reform and softening his demand for limits on litigation in a patient's bill of rights. But these were political calculations and legislative necessities, ways to grease the possible enactment of laws that fulfilled his pledges. His new attitude toward nation building also arose in part from need: it was what allies wanted, and Bush wanted allies to be happy. But it also reflected a deeper grappling with the issue, an openness toward changing circumstances and the changes they required from him. Some advisers had nudged him, but Bush had the final say. And Bush was saying something different from what he had at the start.

He was acting differently, too, but only by degrees. He maintained certain rituals and routines, along with his capacity for detachment. As I watched him and received reports about him through September and October, I was struck by the way he at once held his sorrow and burden close and kept them at bay, his days an amalgam of the extraordinary and the ordinary, of new obligations and old habits.

Several aides who flew into Manhattan with him aboard

Marine One in early October remarked on the somber hush that fell over Bush and the others in the helicopter as it swept near the devastated site of the World Trade Center. One of Bush's best friends, Don Evans, the commerce secretary, was sitting with him, as were Rove and others whom Bush would usually engage in animated conversation. It was, as one of the aides who accompanied them told me, "a joke-y crowd" under usual circumstances. But that was not the case on this flight. Once Marine One pulled within sight of lower Manhattan, no one said a word. The silence stretched on and on, and when Bush turned away from the window, his eyes were wet with the beginnings of tears.

Soon enough, however, he turned typically merry and mischievous. At one of the events he attended, he publicly ribbed Mayor Rudy Giuliani about Giuliani's hopes for the Yankees to make the World Series. (The Yankees ultimately did.) He teasingly reassured a New York City schoolgirl who said that her favorite class was the dead space between classes that she seemed like a pretty normal teenager to him. He beamed and bantered, just like before.

In the middle of White House meetings or in impromptu comments to advisers, Bush would lend expression to grave thoughts, saying that "this is war" or that "things have changed forever," as if repeating the phrases that played endlessly on a repeating tape loop in his mind. But in chats with siblings, he would ask after nieces and nephews, his inquiries as routine as ever. "George has not changed in my conversations with him," said Jeb Bush in an e-mail response to several questions that I had sent him. Jeb Bush said that the president seemed "relaxed and comfortable" and that he could "still talk about the family things—how the children are doing, for example—under this enormous pressure."

For more than a month after the attacks, with only one or two exceptions, each of Bush's meetings focused exclusively on what the nation needed to do, both abroad and at home, to strike back against the perpetrators, cope with the threats to its security and rebound from the damage, physical and economic. His push for education reform, his desire to give more government social service money to religious groups—these receded from his schedule and his attention, matters that suddenly seemed minor. Now he was sometimes coming into the West Wing shortly before 7:00 A.M. instead of shortly after and entering into phone conversations with world leaders and tight clutches with congressional leaders more frequently than ever before.

But the general architecture of his days and even his weeks was unchanged. He still went to bed between 9:30 and 10:00 P.M. and still woke up at 5:30 A.M., getting his usual seven and a half hours of sleep. (One family friend told me that Bush's father had often spoken about not getting enough rest and pacing himself during the Persian Gulf War and that the second President Bush was remembering and heeding those words.) He still took time at lunch or after he left the Oval Office around 6:00 P.M. to work out on the treadmill and weight machines in the White House and still took long runs at Camp David, to which he continued to travel almost every weekend, still leaving the White House midway through Friday afternoons. He still invited friends to pass the time with him there, just as he still had friends to dinner at the White House during weeknights. On the weekend in early October when he gave final authorization for the military strikes, he made sure to tell Hughes, who was at Camp David with him, to bring her husband and to tell Card, who was also there, to bring his wife. As the pilots got ready for the first air raids over Afghanistan to begin on Sunday, Bush watched some of the Saturday football game that the University of Texas was playing.

His nighttime reading did not appear to be focused on the

region of the world that was occupying his daytime considerations. White House aides said that in the weeks immediately following September 11, he worked his way through two books that were indeed about military engagements that more or less affected the course and future of the country but were true to the kinds of history books he had long preferred. One was *April 1865,* by Jay Winik, about the Civil War. The other was *Ghost Soldiers,* by Hampton Sides, about a daring World War II rescue. One aide told me that Bush was also leafing through a novel, *Code of the West,* by Aaron Latham, that recast the Arthurian legend in terms of the Wild West of Texas in the 1860s, where characters uttered pleasantries like "thanky, ma'am" and Excalibur was transmogrified into a Colt revolver. How perfect for Bush, who had often transformed statesmanship into something with less elegant cadences and shorter, blunter, homier words. How fitting a metaphor for his manner of tweaking the office of the presidency so that it conformed to his own sensibilities.

So while Bush seemed different, he also seemed the same, articulating the exceptional while hewing to the usual, reaching high while trying to stay grounded, mingling urgency with normalcy. That may well have been exactly what he and his advisers wanted and intended to project, their chosen mood and image for a country that needed to be both rallied and consoled, exhorted and soothed. But it was also utterly Bush, whose attachment to his ways and sporadic detachment from his worries came to look like more than just interesting personality quirks. Perhaps they were also useful methods for maintaining stability during unstable times. And perhaps his resistance to letting himself be consumed by his obligations also prevented him from being overwhelmed by them.

Epilogue

IN NOVEMBER, DURING THE HOLY MONTH OF Ramadan, Bush hosted a state dinner for fifty ambassadors from Muslim countries that featured traditional Arab foods. This from a guy who cherished his peanut butter and jelly and whose grasp of the Muslim world—really of all geopolitics—had been such a question mark that he left the impression, just two years earlier, that he thought the Taliban was a musical group.

Bush also stood beside General Pervez Musharraf, the president of Pakistan, at a news conference in New York City to announce $1 billion in American aid to Afghanistan's pivotal neighbor. This from a guy who, in that pop quiz on his way to the White House, had been unable to name the leader of the country he was now trying to make America's new best friend.

All the while Bush was amassing a degree of executive power that, in the eyes of some scholars, rivaled President Franklin Delano

Roosevelt's during World War II. He signed an executive order allowing him to request secret military tribunals for accused terrorists; he announced dramatic reductions in the U.S. nuclear arsenal and, by not putting the cuts in a treaty, avoided a vote by the Senate to ratify the decision. This from a guy who, on taking office, had seemed intent on making the presidency a smaller, humbler thing.

It was all somewhat difficult to process. Until September 11, Bush had never cut a profoundly commanding political figure; his pursuit and attainment of the presidency had been less a passionate quest than an excellent adventure. He lacked the churning ambition of Lyndon B. Johnson, the roiling demons of Richard Nixon, the pristine idealism of Jimmy Carter, the ideological certainty of Ronald Reagan, the enormous and self-destructive appetites of Bill Clinton. He was simpler and plainer than most of his predecessors, and that seemed to be one reason many Americans voted for him: he was an unthreatening, easygoing man for unthreatening, easygoing times. Yet the abruptness with which those times had changed, trapping Bush in a labyrinth of obligations for which he was ambiguously prepared, was turning him into one of the most interesting presidents in decades. He had indeed exceeded expectations, at least as an unusually riveting character in a newly compelling drama.

He was prosecuting a "war," for lack of any better word, unlike any before it. It had no discernible geographical boundaries. It existed in America, where terrorist cells might still be hatching plots and letters laced with anthrax forced members of Congress to flee their offices. It required intelligence operations in Europe and South America, where those cells might also exist. It played out in the Middle East, where the grudges of Muslim extremists flared most brilliantly and chillingly. It had an almost impossible goal—the elimination of terrorists with a global reach—and no conceivable timetable.

He was severely curtailing certain civil liberties, sparking a debate that promised to grow in fury and complexity. In the name of investigating and thwarting potential terrorist activity, hundreds of Middle Eastern men in the United States were being detained and, in the end, sometimes prosecuted on minor violations that had nothing to do with terrorism and would not otherwise have come to light. The Justice Department in mid-November asked law enforcement officials across the country to pick up five thousand men, mostly from Middle Eastern countries, for questioning, even though these men had entered the country legally over the previous two years. Federal officials had new, broader authorities for wiretaps and a new, broader ability to monitor conversations between lawyers and some people in federal custody. More was sure to come.

On top of all of this, Bush was confronting a persistently sluggish economy that refused to grow, and presiding over a serious rift between Republicans, who controlled the House, and Democrats, who controlled the Senate, over how to coax it along. The Republicans wanted more tax cuts, particularly for businesses, while the Democrats wanted more federal spending and extra support for people lower down the economic ladder. For Bush, choosing the correct course might very well determine whether he lived the latter years of his presidency in acclaim or shame.

There was one sense in which all this vindicated him. I could not help but remember back to all the times when he answered questions about his fitness for the presidency, in terms of his policy expertise and body of knowledge, by saying that the more relevant test was whether he had the judgment and prudence to put the right advisers beside him. The country's problems and challenges as the first year of Bush's presidency came to a close were undeniably too complicated for one man, and Bush had indeed surrounded himself with some seasoned players who at times provided reasons for confidence. Rumsfeld and Powell, for instance, seemed to have a handle on their jobs. Despite an approach to some domestic matters, like the environment, that could give a person serious pause, Cheney certainly added something to the combined wisdom of Bush's foreign policy advisers. But was O'Neill, the treasury secretary, who often sang from a different songbook than almost everybody around him, on top of the economy? And what of Ashcroft? Could he be trusted to protect all Americans' rights?

Former Wisconsin governor Tommy Thompson, the secretary of health and human services, initially dismissed the first anthrax case, in Boca Raton, Florida, as the possible result of something its victim had picked up from a polluted stream on a vacation in North Carolina. Thompson was also hard pressed to explain the subsequent anthrax-related deaths of two postal workers in Washington, D.C. They had worked in the postal station that handled the tainted mail that ended up in the office of Tom Daschle, the Senate majority leader, but while uninfected Senate aides were lavished with medical attention, federal officials did not think to worry about mail handlers.

Former Pennsylvania governor Tom Ridge came on board as the head of a new Office of Homeland Security and proceeded to hold a string of news conferences at which he was unable to answer even basic questions about the new tasks before him. If an observer factored in the humiliation of former New Jersey governor Christie Whitman, the head of the Environmental Protection Agency, as Bush balked at protecting the environment, it seemed fair to conclude that there was no worse fate for a former Republican governor than ending up in Bush's cabinet. Oklahoma governor Frank Keating, once in contention for the vice presidency or a top administration job, must have been thanking his lucky stars.

Bush tasted some real and impressive success. Little more than a month after the air strikes in Afghanistan began, the Taliban surrendered its control of Kabul, the Afghan capital, which followed the Taliban's rapid-fire defeat in other cities. The brisk pace of America's military accomplishments stunned even those officials directing the effort, and Bush was suddenly facing the conundrum of assembling a new government in Afghanistan before the United States was really ready.

He was also facing trickier decisions about what would come next. In late November, he turned his attention toward Iraq, another country believed to sponsor terrorism against the United States. The administration began making noises about forcing Iraq to open its borders to United Nations inspectors so that it could be determined whether Saddam Hussein was producing weapons of mass destruction. This was what hawks and some of Bush's conservative supporters wanted, but it worried many European and Arab countries in the loose coalition that Bush had assembled. Afghanistan and the Taliban had been easy sells: allies understood the imperative for Bush to avenge the deaths of so many Americans on September 11. The rest of it would be much, much harder.

And Bush would have to shoulder it, supervising all his high-powered advisers and aides, giving them their orders, filtering their counsel, resolving their differences. There was nothing small or humble about that, and Bush would have to be larger than he had ever seemed poised to be—larger than even some of his supporters had ever perceived him. Was he? In the weeks after September 11, when most Americans wanted to believe the best, I certainly came to wonder if Bush's many critics had sold him short, and I came to wonder if *I* had. It was ultimately unknowable in the present; only the hindsight of history, and the eventual weakening and permeation of the wall that the White House had set up between Bush and those of us chronicling him, would yield something closer to the truth. Until then I was left with a motley and inconclusive collection of stray details that might or might not be relevant. I rifled through them constantly for a better sense of what sort of president we had in Bush.

We had a President Bush who, to a significant degree, was what Candidate Bush had augured. For all my qualms about the nuttiness of a presidential campaign, I saw that the product Americans had purchased often matched the prior advertisement of it, or at least the advertisement as reporters registered and rendered it, pruned of fictive embellishments, trimmed of fat. The Bush we got was the Bush we glimpsed: irreverent to a fault, with a tongue like a Slinky toy and a temperament like an easily wowed—and just as easily tired—little boy's. The Bush we got was happy to be his own tenuously engaged self, just as he had always been, except when it seemed like the public might be looking for—and the presidency might be served by—something bigger. Then the Bush we got was the South-Carolina-town-hall-meetings Bush, transformed by his own extra effort and by his handlers' packaging into a more active, vibrant, probing leader.

He was palpably sensitive, at least on an emotional front—capable of real feeling, and of being moved and motivated by it. There were stories aplenty after September 11 of his poignant encounters with people touched and diminished by that horrific day, and some of these people spoke of the comfort they took from Bush or the strength they saw in him. He was humorously bumbling; I could not get out of my head the tales that someone in the White House told me about how much trouble Bush had with the simple chore of reading and recording his radio address. He would flub key words and lines, grow flustered and irritable and have to start over, sometimes more than once. Did that matter, or touch on something more significant? Probably not.

He was fetchingly down-to-earth. On the last day of November, as he boarded Marine One to head to Camp David, he walked past the uniformed officer who always stands beside the steps to the helicopter without offering the officer the customary salute. Maybe it slipped Bush's mind; maybe he just didn't have a free hand, because he was cradling the unruly mop of black hair that was his dog Barney. But after climbing to the top of the steps, Bush placed Barney in the helicopter, turned around and went back down the stairs. At this point the unsuspecting officer's back was to Bush; the president had to tap him on the back several times so that he turned around. And then Bush smiled and saluted.

Was he also deeper and more curious than he usually let on? Perhaps. I recalled a long-ago conversation with him that contradicted my onetime insinuation in at least one, and maybe more, of the articles I had written that he wasn't much of a reader. Bush mentioned, in response to a question, that he had just finished *One Nation, Two Cultures,* by Gertrude Himmelfarb, a serious intellectual examination of the cultural divide in American society following the 1960s. There was no way he pulled that title out of the air. I also flashed on a conversation with Bush's buddy Joe O'Neill about the deliberately diverse weekend groups that the president would put together at Camp David. O'Neill said that the president's Texas friends and top White House officials and lowest-tier White House aides and a cabinet member would sit at a table for long, discussion-dense meals. Without seeming for an instant to be peddling the point, O'Neill made it sound like something of an egalitarian salon.

I nonetheless expected to look up at the television set toward the end of 2001, after so many months of national anxiety, and see, in the frame, a Bush who looked 150 years old, slumped and bowed, wavering and quivering in the eye of a maelstrom. This would not have been reassuring, but I'm not sure it was reassuring, either, to look up in late October and see, in the frame, Bush looking strangely jocular as he said, several times, "I don't have anthrax." The substance of the scene, as Bush talked briefly with a handful of pool reporters in the Cabinet Room, would have been disturbing in and of itself: an American president was making clear to the country that he himself had not been a victim of bioterrorism. But Bush's manner made it more unsettling. He was dodging questions about whether he had been tested for exposure to anthrax by simply and coyly saying he was not infected, and it had the air of a silly game. It felt wrong, with Bush's demeanor yet again at odds with the gravity of the circumstances.

Some things never changed, his loose-y goose-y bearing among them. When Putin visited him in Crawford and they held a sort of town hall meeting with students at the Crawford High School on November 15, he seemed to flop around at his lectern, slouching over it, leaning far to one side, then far to the other, his feet constantly shifting and shuffling. It was peculiar enough that one of the junior White House aides later entertained reporters with an imitation of what Bush's legs must have looked like behind the lectern, which amounted to a spastic slow-motion dance.

At one point one of the students asked Bush how he thought the fall of the Taliban government might affect Afghan women's rights. Bush basically said nothing concrete, tossing the question to Putin, who did. Putin noted that the persecution of—and discrimination against—women was a worldwide issue that had "taken extreme form" in Afghanistan. "In many countries of the world," he said, "especially in poor countries, this problem exists and has acquired dramatic dimensions. To overcome this, one needs to develop specific gender-oriented programs that would include, primarily and first of all, questions related to proper education for women. And I would like to reiterate, there are many programs and many people devoted to implementing such specific, special activities for the benefit of women."

Here is what Bush had said, sharing a memory from his first meeting with Putin in June: "So we are getting ready to have the first press conference we had together in Slovenia. And, by the way, there was, I think, a thousand reporters there—it seemed like a thousand. And we were walking in. I said, 'Say, I understand you've got two daughters.' He said, 'Yes.' He said, 'They're teenagers.' I said, 'I've been through that myself.' I said, 'Who did you name them for?' He said, 'Well, we named them for our mothers, my mother and my mother-in-law.' I said, 'That's interesting—that's exactly what Laura and I did, too. We named our girls for our—my mother and Laura's mom.' And I said, 'Gosh, the thing I want most in life is for those girls to be able to grow up in a free world and prosper and realize their dreams.' He said, 'That's exactly what I hope for, as well.' There's a lot in common, even though—between our countries, even though it's a long way away. And it all starts with the human element, the thing that matters most in life, and that is our faiths and our families and our respect and loves as dads for our daughters."

Evasive, mundane, stuttering, sincere, sweet and, in the end, both a bit platitudinous and a bit profound—it was all there, and it was all Bush. When another student asked him for a little advice, Bush said, "You never know where life is going to take you. I can assure you, when I was a senior in high school, I never sat in an audience saying, 'Gosh, if I work hard I'll be president of the United States.' Didn't exactly fit into my vocabulary in those days. But you never know. You never know."

Almost everyone laughed and almost everyone applauded.

Acknowledgments

THERE WERE LITERALLY SCORES OF PEOPLE WHO in some way helped with this book, or who helped me get through the long days and long weeks of covering the 2000 presidential campaign and the White House. It would take pages and pages to name them all, so I beg the forgiveness of those not mentioned here.

I owe an enormous debt to John Berman, Patsy Wilson, Anne Kornblut, John Dickerson and Campbell Brown for their generosity as colleagues and support and encouragement as friends, in terms of this book and so much else. The next round is on me.

Terry Neal, Jackie Calmes, Glen Johnson, Richard Wolffe, Kevin Flower, Candy Crowley and Mike Roselli also provided great company and comfort on the road. I would hop on a bus or plane with them any time, though I would hope it was not traveling toward a political rally.

Thanks to so many people at the *New York Times* for giving me either the opportunities I benefited from, the confidence I needed, the breathing space I requested or the advice, reassurance and friendly conversation I frequently sought. They include, but are by no means limited to, Jim Roberts, Renee Murawski, Kit Seelye, Maureen Dowd, Don Van Natta, Lizette Alvarez, Vivian Toy, Carey Goldberg, Alison Mitchell, Frank Clines, Jim Dao, Adam Nagourney, Monica Borkowski, Dick Stevenson, Jim Yardley, Kate Phillips, Janet Elder, Adam Moss, Jill Abramson, Rick Berke, John Broder, Linda Lee, Jan Battaile, Joyce Purnick, Mike Oreskes, Andy Rosenthal, Gerald Boyd, Bill Keller, Joe Lelyveld and Howell Raines.

Thanks to Lisa Bankoff for coaxing me, David Hirshey for getting me across the finish line and Alison Callahan for providing crucial continuity and counsel. Thanks to Robert Jones for all the improvements—for being such a brilliant editor and such a loyal, special friend. He was one of a very rare kind, and the faithful, adoring band of us who were privileged to call ourselves "Robert's authors" will always miss and never forget him.

Thanks to Laura Nichols for humoring me and perking me up. Thanks to Elinor Burkett for propping me up, keeping me steady and watching over me, always and forever. Thanks to Dennis Gaboury for making sure my home never became a hovel.

Thanks to my father, Frank, for refuge at a crucial time. Thanks to my sister, Adelle, and her husband, Tom, for the regular phone calls—and all the laughs.

Thanks, finally, to my sources, often unnamed but never unappreciated. I think—I hope—they know that.

Notes on Sources

THE VAST MAJORITY OF THE INFORMATION IN this book comes from my own observations, experiences and reporting, including hundreds of interviews, over the course of two years, with George W. Bush's aides, friends and relatives; with politicians, academics, political experts and political operatives; and with Bush himself. Many anecdotes in the text are not, for the sake of brevity or discretion, given a specific source—they were usually shared with me by people with reliable knowledge of what happened. Other anecdotes and information were so widely reported, by so many different journalists, that there is no single source that warrants attribution. What follows are the exceptions to the above—examples of specific pieces of material from which I borrowed or to which I felt a debt.

In the prologue, the narrative details of Bush's hours and days between September 11 and September 20 were supplemented by stories by David E. Sanger and Don Van Natta Jr. in the *New York Times* on September 16, 2001, and by Howard Fineman, T. Trent Gegax, Debra Rosenberg and Matt Bai in *Newsweek* on September 24, 2001.

In chapter 1, some of the descriptions of Bush's childhood and school years were supplemented by stories by Lois Romano and George Lardner Jr. in the *Washington Post* in July of 1999 and by Nicholas D. Kristof in the *New York Times* in May and June of 2000. These descriptions were also supplemented by passages of the book *First Son: George W. Bush and the Bush Family Dynasty,* by Bill Minutaglio, published in 1999 by Times Books/Random House.

In chapter 2, Bush's statement that he "never interviewed" the journalist who theorized that he might have dyslexia is rendered differently than it was in most news accounts at the time, including my own in the *New York Times*. Those news reports relied on an oral "pool report" from a Fox News producer who was in charge of the only camera crew on the scene and who unintentionally provided a slightly altered version of Bush's actual quote. When I went back to review the episode, I noticed that a Fox News transcript from September 14, 2000, includes the accurate quote, which is the one used here.

Also in chapter 2, some details and quotes regarding the debate between George H. W. Bush and Geraldine Ferraro, along with parts of the characterization of the elder Bush's attitude toward debates, come from a story by R. W. Apple Jr. in the *New York Times* on September 24, 2000. The account of Karl Rove saying, in a post-election forum, that he wished he had not let George W. Bush skip the first Republican debate during the buildup to the primaries comes from a story by Richard L. Berke in the *New York Times* on February 12, 2001.

In addition, chapter 2's characterization of Al Gore's academic record at Harvard comes from a story by David Maraniss and Ellen Nakashima in the *Washington Post* on March 19, 2000. Bill Bradley's verbal SAT scores were first reported in *Slate* and later reported in several places, including a brief article in *USA Today* on January 27, 2000. The reference to John McCain's academic record at the U.S. Naval Academy also comes from a variety of sources, including a story by Amy Argetsinger in the *Washington Post* on March 6, 2000.

In chapter 3, the detail about Bush taking twenty-four-hour leaves from his father's campaign and flying home from the West Coast to Washington, D.C., comes from a story by Nicholas D. Kristof in the *New York Times* on August 29, 2000. The detail about Bush telling David Dreier that a good way to raise money was to mine a mother's Christmas-card list for names comes from a story by Evan Thomas and Martha Brant in *Newsweek* on August 7, 2000. The detail about Hughes asking Bush if she had to "grovel" for forgiveness from me comes from a story by Evan Thomas and Martha Brant in *Newsweek* on February 28, 2000. Some of the details about the way allies of the Bush campaign waged war against John McCain in South Carolina come from a story by Eric Pooley, James Carney, John F. Dickerson and Maggie Sieger in *Time* on February 28, 2000.

In chapter 4, the analysis of the first batch of judges whom Bush nominated for the federal bench was supplemented by stories by Elisabeth Frater in the *National Journal* on May 19, 2001 and by the staff of *The Nation* on June 4, 2001.

In chapter 5, the detail about Bush grumbling that he had to "feed the beast" by going back in his plane to talk to the reporters there comes from a story by the *Newsweek* staff in *Newsweek* on November 20, 2000. The details about the way Kent Hance campaigned against George W. Bush in 1978 and what Bush learned from that campaign come from a story by Nicholas D. Kristof in the *New York Times* on July 27, 2000. The details about Bush's fondness for the Broadway musical *Cats,* the movie *Cool Hand Luke* and Julie Christie in *Doctor Zhivago* come from a column by Maureen Dowd in the *New York Times* on October 3, 1999.

In chapter 6, the detail about Bush challenging his father to go "mano a mano" was apparently first reported in a story by David Maraniss in the *Washington Post Magazine* on January 22, 1989.

In chapter 7, the portrait of Laura Bush and her relationship with her husband was supplemented by stories by Helen Thorpe in *George* in April 2000; Nicholas D. Kristof in the *New York Times* on July 29, 2000; by Elaine Sciolino in the *New York Times* on January 15, 2001; and by Ann Gerhart in the *Washington Post* on March 22, 2001.

In chapter 8, the account of how Bush prepared for his August convention speech and general election debates was supplemented by a story by Michael Duffy, James Carney, John F. Dickerson, Tamala M. Edwards and Karen Tumulty in *Time* on November 20, 2000.

In chapter 11, the analysis of Bush's evolving position on "nation building" was supplemented and informed by a story by David E. Sanger in the *New York Times* on October 12, 2001.